Hugo Esteban Rios

3º A

AF892249

Mathematics 3º ESO

(Adaptado a la LOMCE)

Mathematics 3º ESO

Adaptado a la LOMCE

Autor
Javier Sánchez Pina

ISBN
978-84-608-9759-0

Depósito Legal
MU-811-2016

1ª Edición
Julio de 2016

ÍNDICE

Nota del autor

This material contains enough theoretical development and proposed exercises with solutions, in order to promote our students' autonomy. Moreover, for almost each introduced concept, some exercises are solved step by step.

To prepare this material, I have used the following bibliography:

Bibliography	- Fundamentals of Mathematics (Authors: J. Van Dyke, J. Rogers, H. Adams; Ed.: BROOKS/COLE, CENGAGE Learning), - Spanish text books: - Matemáticas 3º ESO (Autores: J. Cólera y otros; Ed.: Anaya). - Matemáticas 3º ESO (Autores: A. M. Gaztelu y A. González; Ed.: Santillana). - Notes and work sheets (in English and Spanish): Alfonso González, José A. Jiménez Nieto, Miguel Ángel Hernández Lorenzo, Isabel Torres (IES Ruradia), IES Don Bosco. - English Dictionary (Oxford Edition).

Esta edición del libro tiene muchísimos menos errores tipográficos, de idioma y de cálculo gracias al profesor **D. Mariano Salido Esparza**. Mi más sincero agradecimiento. Profesores con esta dedicación hacen grande esta profesión.

This materials have been reviewed in order to fulfil new Spanish **LOMCE** regulations.

Para el profesor: <u>Evaluación por estándares:</u> He preparado un cuadro en el que relaciono cada estándar evaluable de la LOMCE con los ejercicios de este libro que lo trabajan, y que puede ser muy útil a la hora de diseñar las pruebas de evaluación. Si estás interesado en que te lo envíe, ponte en contacto conmigo en el mail de más abajo.

En mi web (www.cuadernodepitagoras.com) he subido las presentaciones que uso con mis alumnos en clase. Si en algo crees que te puedo ser de utilidad, si encuentras erratas, o si deseas hacerme alguna sugerencia, te agradeceré que te pongas en contacto conmigo en el mail javiersanchezpi@gmail.com.

Soy doctor en Química y licenciado en Ciencias Ambientales y soy profesor de matemáticas en un instituto en Murcia (España).

Espero te sea de utilidad y que contactes conmigo para lo que precises.

Un saludo. Javier Sánchez Pina.

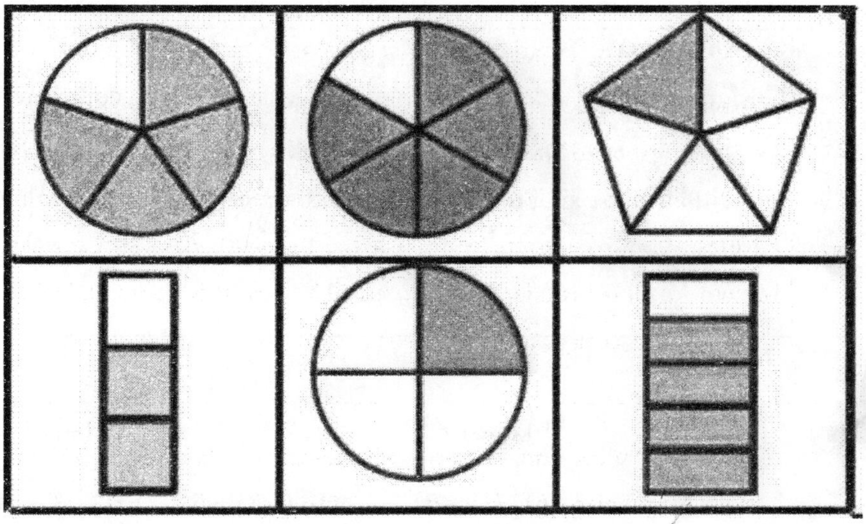

Unit 1.- Rational numbers

1. Integers

Natural numbers

The set of Natural Numbers (they are also called *whole* numbers) is $\mathbb{N}=\{0,1,2,3,4...\}$.

Natural numbers have two purposes:

– Counting: “There are three apples on the table” → CARDINAL NUMBERS

– Ordering: “Barcelone is the second largest city in our country”→ORDINAL NUMBERS

Integers

Negative numbers are an extremely useful tool for many kinds of problems. For instance, the concept of a negative temperature, the notion of a negative balance in a bank account, and the use of negative numbers to describe the motion of an object in a particular direction are important uses of negative numbers.

Addition and subtraction of two integers

To add two integers

1. If the signs are ***alike***, ***add*** their absolute values and use the common sign for the sum.

2. If the signs are ***not alike***, ***subtract*** the smaller absolute value from the larger absolute value. The sum will have the sign of the number with the larger absolute value.

Exercises

1. Add:

a) 1 + 3 b) (– 8) + 5 c) 2 + 1 d) (–4) + 4 e) 5 + (- 6)

f) (–6) + 4 g) 7+ (– 2) h) (– 3) + 5 i) (– 3) + (– 2) j) (– 4) + (– 2)

k) (– 2) + (– 2) l) (– 5) + 6 m) (– 2) +(– 5) n) (– 5) + 10

ñ) (– 7) + (– 4) o) (– 4) + 7 p) (-5) + (– 2) q) (- 7) + (– 10)

2. Complete:

a) (-11) + [....] = 4 b) (+13) + [....] = 12 c) [....] + (-20) = (-12)

d) (+3) + [....] = (-7) e) (-15) + [....] = 9 f) [....] + 8 = 7

Addition of more than two integers

To calculate additions and subtractions implying more than two numbers:

- 1st: add separately positive numbers,
- 2nd: add separately negative numbers,
- 3rd: finally, calculate the difference:

TOTAL POSITIVES – TOTAL NEGATIVES

Example: Calculate:

a) $7 - 2 + 4 + 1 - 8$ b) $10 + 2 - 5 - 9 + 2$ c) $-2 + 5 - 4 - 10 + 1$

Solution:

a) $5 - 2 + 4 + 1 - 8$

1st: add separately positive numbers → $+7 + 4 + 1 = 12$

2nd: add separately negative numbers → $-2 - 8 = -10$

3rd: finally, calculate the difference → $12 - 10 = \boxed{2}$

b) $10 + 2 - 5 - 9 + 2 = 14 - 14 = \boxed{0}$

c) $-2 + 5 - 4 - 10 + 1 = 6 - 16 = \boxed{(-10)}$

No more than one sign (+) or (-) before a number!!!

In Maths, it is no allowed to write two signs (+) or (-), together before a number. When this happens, we must change both signs by only one. We will do it as follows:

You must pay attention to the first sign:

- If the first sign is ***(+)***, then, quit it and ***leave*** the second one.
- If the first sign is ***(-)***, then, quit it and ***change*** the second by its opposite.

Example:

a) $+(-7) = (-7)$ b) $-(-2) = +2$ c) $-(+8) = (-8)$ d) $-(+1) = (-1)$

e) $+(+11) = +11$ f) $+(-14) = (-14)$

Exercises

3. Quit parentheses and work out:

a) (+3) – (+8) b) (–9) + (–6) c) (–7) – (–7) – (+7)

d) (–11) + (+8) – (–6) e) (+15) – (– 12) – (+11) + (–16)

f) (–3) – (–2) – (+4) + (–7) + (+8) g) (+11) – (+7) + (–13) – (–20) + (–11)

4. Quit parentheses and work out:

a) -(- 3) + (-7) - (- 2) + (- 9) b) 3 - 7 - (- 2) + 9 c) 15 + (- 21) – 20 – (- 1

d) (- 52) - 24 e) (- 72) + 80 – 8 f) 2 + (-3) + (-5) + (-4) + 6

g) (-3) + 6 + (-4) + 4 + (-6) h) (- 1) - 6 + (- 7) - (- 5) - 21 + (- 12)

Product and division of integers

Product and division of integers follow identical rules:
- Product or division of integers of the **same sign** will give a **positive number**.
- Product or division of integers with **opposite sign** will give a **negative number.**

Exercises

5. Calculate the following products:

a) (+21) · (+3) · (+4) b) (+19) · (-2) · (+3)

c) (+13) · (-5) · (-6) d) (-20) · (-9) · (-3)

6. Calculate the following products:

a) (-5) · ☐ = (-30) b) ☐ · (+3) = 45 c) (-9) · ☐ = 27

d) ☐ · (-8) = (-48)

Exercises

7. Calculate the following quotients:

a) 25 : (– 5) b) (– 21) : (– 7) c) 8 : (– 4) d) 4 : (– 1)

e) 8 : (– 4) f) (– 24) : (– 3) g) (+ 12) : 2 h) 72 : (– 8)

8. Copy and complete:

a) (-36) : □ = (- 4) b) (-54) : □ = 9 c) □ : (-6) = (-42)

d) 48 : □ = (- 6) e) (-63) : □ = (-7) f) □ : 8 = 2

Order of operations

To evaluate an expression with more than one operation:

- Step 1. Parentheses
- Step 2. Exponents
- Step 3. Multiply and Divide
- Step 4. Add and Subtract

Example: Calculate: a) $1 - 5 \times 4$ b) $3 + 4 : 2$ c) $1 + 4 \times (3 + 2)$

d) $5 \times 4 - (1 + 2)$ e) $18 - [(3 + 6 + 9) : (9 - 6)]$

f) $[(55 - 10) - (3 \times 6 \times 9)] : (-3)$

Solution:

a) $1 - 5 \times 4 = 1 - 20 = \boxed{(-19)}$.

b) $3 + 4 : 2 = 3 + 2 = \boxed{5}$

c) $1 + 4 \times (3 + 2) = 1 + 4 \times 5 = 1 + 20 = \boxed{21}$.

d) $5 \times 4 - (1 + 2) = 20 - 3 = \boxed{17}$.

e) $18 - [(3 + 6 + 9) : (9 - 6)] = 18 - [18 : 3] = 18 - 6 = \boxed{12}$

f) $[(55 - 10) - (3 \times 6 \times 9)] : (-3) = [45 - 162] : (-3) = (-117) : (-3) = \boxed{39}$

Example: Calculate: $7 \times 5 - [-(8 - 3) + (12 - 6)] + 5 + 4^2 : 8$

Solution:

$$7 \times 5 - [-(8 - 3) + (12 - 6)] + 5 + 4^2 : 8 =$$

Parentheses: $= 7 \times 5 - [\ -5 \ + \ 6] \ + 5 + 4^2 : 8 =$

Brackets: $= 7 \times 5 - \ 1 \ + 5 + 4^2 : 8 =$

Powers: $= 7 \times 5 - \ 1 \ + 5 + 16 : 8 =$

Products: $= \ 35 \ - \ 1 \ + 5 + 16 : 8 =$

Quotients: $= \ 35 \ - \ 1 \ + 5 + \ 2 \ =$

Sums and subtracts: $= \ 42 - 1 = \boxed{41}$.

Exercises

9. Calculate. Pay attention to the order of operations:

a) $35 + 7 \cdot (6 - 11)$

b) $(6 + 2 - 9 - 15) : (7 - 12 + 3 - 6)$

c) $-(8 + 3 - 10) \cdot [(5 - 7) : (13 - 15)]$

10. Calculate. Pay attention to order of operations:

a) $(-3) \cdot [(-9) - (-7)]$

b) $28 : [(-4) + (-3)]$

c) $[(-9) - (+6)] : (-5)$

d) $(-11) - (-2) \cdot [15 - (+11)]$

e) $(+5) - (-18) : [(+9) - (+15)]$

f) $(-4) \cdot [(-6) - (-8)] - (+3) \cdot [(-11) + (+7)]$

11. Work out:

a) $-5 + 4 \cdot (-2 + 1)^3 - (-9 + 6)^2$

b) $-6 - 2 \cdot [-4 + 5 : (-1)]$

c) $12 - 2 \cdot [25 : (-4 - 1) + (-2) - (6 - 10)]$

d) $-7 - (-3) + (-8) \cdot (-1) - (-12) : (-4)$

e) $(-1)^4 - (-2)^3 + 18 : (-9) - (-4 + 2)$

f) $(-5 - 4) \cdot (-2) + 28 : (-7) + (-2)^3$

g) $-5 - 4 \cdot [-8 : 2 - 2 \cdot (-3)]$

h) $6 - 5 \cdot [-4 - 1 + (-2)^2 - 32]$

i) $12 - 8 \cdot [-2 + 4 : (-1) - (-3 + 2)^4]$

j) $(-2)^5 : (3 + 1)^2 + 2 \cdot (-5 - 4 + 3)$

k) $10 - 10 \cdot [-6 + 5 \cdot (-4 + 7 - 3)]$

2. Fractions

Integers are useful in counting or ordering elements, but not in expressing measurements. When measuring, it is usually necessary to fractionate the unit: half, four thirds, seven tenths… These measurements are expressed by using fractions:

$$\frac{1}{2} \quad \frac{4}{3} \quad \frac{7}{10}$$

A ***fraction***, $\frac{m}{n}$, is the indicated quotient of two integer numbers, where divisor is not zero. It is interpreted in the following way:

· ***n*** expresses the number of portions made from one unit.

· ***m*** expresses the number of portions that are taken.

Dividend is named ***numerator***, and divisor, ***denominator.***

Remember how to read fractions. The method to name the denominator of a fraction depends on the denominator:

- If denominator ≤ 10, we use the ordinals to name it. Example: $\frac{2}{5} \rightarrow$ two fifths, $\frac{9}{6} \rightarrow$ nine sixths.

  There are two exceptions, for denominators 2 and 4: $\frac{1}{2}$ is a half and $\frac{3}{4}$ is three quarters.

- For denominators > 10, we say *"over"* and use the cardinal: Example: $\frac{4}{11} \rightarrow$ four over eleven.

Extension of the concept of fraction

Concept of fraction can also be interpreted as:

a) ***Function or operator:*** For example, when we say: the three fifths of 100 €,

$$\frac{3}{5} \cdot 100 = \frac{3 \cdot 100}{5} = 60 \ euros.$$

b) Proportionality: For example, when we say: three of each four, $\frac{3}{4}$.

c) Percentage: For example, when we say: 20 % of 150 students, $\frac{20}{100}\cdot 150 = \frac{20\cdot 150}{100} = 30$ students.

Exercises	
	12. Find the $\frac{2}{3}$ of 30 and, later, the $\frac{4}{6}$. Why do you obtain the same value? **13.** In my classroom, 3 of each 7 students use glasses. If we are 28 students, how many of us do use glasses? **14.** Calculate: a) The $\frac{5}{8}$ of 40 b) The $\frac{10}{7}$ of 35 c) The $\frac{5}{3}$ of 84 d) The $\frac{2}{3}$ of $\frac{9}{10}$

Simplifying fractions. Simplest form

Simplifying a fraction is the process of renaming it by using a smaller numerator and denominator. A fraction is completely simplified or in its ***simplest form*** when its numerator and denominator have no common factors other than 1. For instance, $\frac{12}{18} = \frac{2}{3}$.

There are two methods available to simplify a fraction.

Method 1: Its useful for small numbers.

We must divide the numerator and denominator by a common factor. Keep dividing until there are no more common factors.

Example 2: Find the simplest form of $\frac{12}{30}$:   Solution: $\frac{12}{30} = \frac{12 \div 2}{30 \div 2} = \frac{6 \div 3}{15 \div 3} = \boxed{\frac{2}{5}}$

NOTE: If numerator and denominator have as last digits zeros, we can simplify all the common zeros above and below. For example, find the simplest form of $\frac{6000}{4500}$:

Solution: $\frac{6000}{4500} = \frac{60\otimes\otimes}{45\otimes\otimes} = \frac{60}{45} = \frac{60 \div 5}{45 \div 5} = \frac{12 \div 3}{9 \div 3} = \boxed{\frac{4}{3}}$

Exercises

15. Express these fractions in the simplest form. (Some may already be in the simplest form)

a) $\frac{5}{10}$ b) $\frac{30}{36}$ c) $\frac{18}{27}$ d) $\frac{48}{84}$ e) $\frac{45}{66}$ f) $\frac{64}{88}$

16. Express, in the simplest form, which fraction corresponds to these situations:

a) In a bag of 90 pens, 15 are blue

b) The number of girls and boys in our class

c) There are 90 pupils of the 270 who come by bus to the school.

Method 2: General method.

We must find the prime factorization of both numerator and denominator. Then, eliminate all common factors, other than 1, in the numerator and the denominator.

Example 3: Find the simplest form of $\frac{120}{100}$:

Solution: First, find the prime factorizations:

$$\left.\begin{matrix} 120 = 2\cdot 2\cdot 2\cdot 3\cdot 5 \\ 100 = 2\cdot 2\cdot 5\cdot 5 \end{matrix}\right\}$$ Then, simplify the common factors *(Do it by yourself, with your pencil!!!)*

$$\left.\begin{matrix} 120 = 2\cdot 2\cdot 2\cdot 3\cdot 5 \\ 100 = 2\cdot 2\cdot 5\cdot 5 \end{matrix}\right\} \Rightarrow \frac{120}{100} = \frac{2\cdot 2\cdot 2\cdot 3\cdot 5}{2\cdot 2\cdot 5\cdot 5} = \frac{2\cdot 3}{5} = \boxed{\frac{6}{5}}$$

NOTE: If numerator and denominator have as last digits zeros, you should simplify them at the beginning.

For example, find the simplest form of $\frac{7500}{5000}$:

Solution: $\frac{7500}{5000} = \frac{75\otimes\otimes}{50\otimes\otimes} = \frac{75}{50} = \left\{\begin{matrix} 75 = 3\cdot 5\cdot 5 \\ 50 = 2\cdot 5\cdot 5 \end{matrix}\right\} = \frac{3\cdot 5\cdot 5}{2\cdot 5\cdot 5} = \boxed{\frac{3}{2}}$

Exercises

17. Simplify the following fractions by using the general method:

a) $\frac{175}{150}$ b) $\frac{48}{60}$ c) $\frac{330}{66}$ d) $\frac{75}{30}$

18. Simplify the following fractions by using the general method:

a) $\frac{66}{55}$ b) $\frac{360}{540}$ c) $\frac{700}{4900}$ d) $\frac{11 \cdot 7 \cdot 2^2 \cdot 3}{11 \cdot 2^3 \cdot 3}$

19. Simplify the following fractions by using the general method:

a) $\frac{126}{210}$ b) $\frac{126}{252}$ c) $\frac{2^3 \cdot 3^2 \cdot 5 \cdot 2^4}{5^2 \cdot 2^5 \cdot 3^2}$

Equivalent fractions

Imagine you are eating pizza with some friends. One of your friends cuts a pizza into 3 pieces and he eats 1 of them.

After that, you cut another pizza into 6 pieces and eat 2 of them. Who has eaten more pizza?

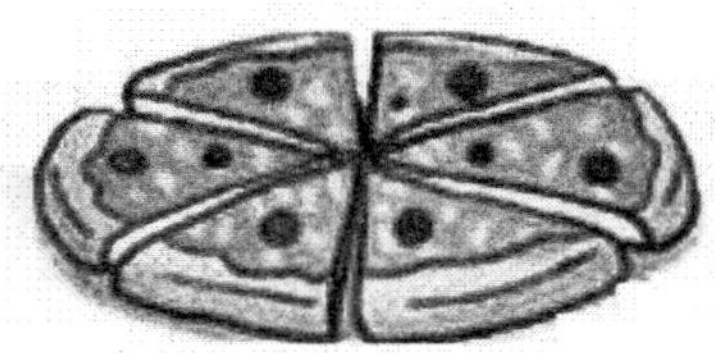

Notice you have eaten the same. These two fractions are ***EQUIVALENT***

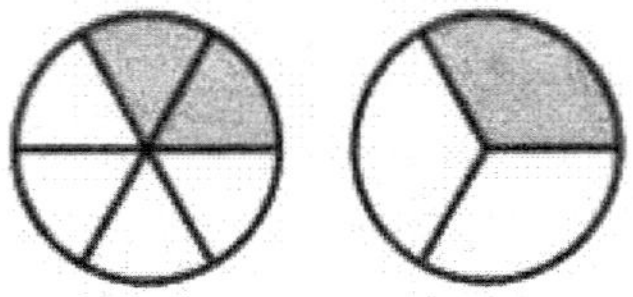

$$\frac{2}{6} = \frac{1}{3}$$

Equivalent fractions are different fractions that name the same amount.

For example:

$$\frac{2}{6}=\frac{1}{3} \qquad\qquad \frac{1}{3}=\frac{4}{12}$$

Equivalent fractions are fractions that have an identical value, even though they look different. Starting with any fractions you like, you can make up a list of equivalent fractions by simply multiplying or dividing both the numerator and the denominator by the same number each time.

Exercises

20. Write four equivalent fractions to the following:

a) $\frac{4}{6}$ b) $\frac{15}{18}$ c) $\frac{14}{21}$

We can check if two fractions are equivalent by cross-multiplying their numerators and denominators. This is also called taking the cross-product.

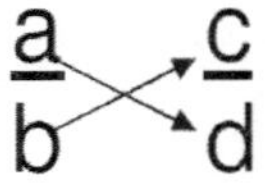

Example: Test if $\frac{12}{20}$ and $\frac{24}{40}$ are equivalent fractions.

Solution:

- 1st cross-product is the product of the first numerator and the second denominator: $12 \cdot 40 = 480$.
- 2nd cross-product is the product of the second numerator and the first denominator: $24 \cdot 20 = 480$.

Since the cross-products are the same, the fractions are equivalent.

Exercises

21. Test if the following pairs of fractions are equivalent:

a) $\frac{2}{3}$ *and* $\frac{10}{15}$ b) $\frac{5}{6}$ *and* $\frac{15}{12}$ c) $\frac{21}{15}$ *and* $\frac{84}{60}$

22. Find out the value of x that makes equivalent each pair of fractions:

a) $\frac{4}{6} = \frac{x}{9}$ b) $\frac{4}{10} = \frac{6}{x}$ c) $\frac{x}{21} = \frac{6}{9}$ d) $\frac{12}{20} = \frac{21}{x}$

Comparing fractions

Imagine you are in a party. You have eaten $\frac{2}{5}$ of a pizza and your friend Andrés has eaten $\frac{4}{7}$ of another pizza. Who has eaten more?

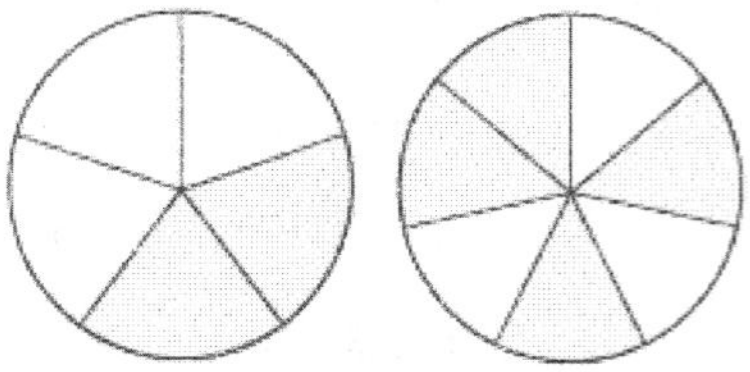

It is hard to answer this question just by looking at the fractions. But, would it be easier if you had had $\frac{2}{5}$ of a pizza and Andrés $\frac{3}{5}$?

What is the difference?   ***THEY HAVE THE SAME DENOMINATORS!!!***

So, first of all, we must transform our original fractions into two new fractions having the same denominator, the Lowest Common Multiple of both denominators.

To ***compare fractions***, first of all, they need to have the ***same denominators***.

This is going to be necessary also when adding or subtracting fractions, so, pay attention to this transformation into Lowest Common Denominator.

Example: Which fraction is bigger, $\frac{2}{5}$ or $\frac{4}{7}$?

Solution:

Step 1: Find out the LCM(5, 7) = 35

Step 2: Transform both original fractions into two new fractions having 35 as denominator.

Remember you can multiply numerator and denominator by the same number:

$\frac{2}{5} = \frac{}{35} \Rightarrow$ To transform 5 into 35, we have multiplied by 7, so, we must also multiply the numerator by 7. So, as, $2 \cdot 7 = 14$, $\frac{2}{5} = \frac{14}{35}$

$\frac{4}{7} = \frac{}{35} \Rightarrow$ To transform 7 into 35, we have multiplied by 5, so, we must also multiply the numerator by 5. So, as, $4 \cdot 5 = 20$, $\frac{4}{7} = \frac{20}{35}$

Step 3: Compare the new fractions: As $\frac{14}{35} < \frac{20}{35} \Rightarrow\Rightarrow\Rightarrow$ $\boxed{\frac{2}{5} < \frac{4}{7}}$

Exercises

23. Which is bigger?

a) $\frac{3}{5}\ or\ \frac{7}{15}$ b) $\frac{3}{7}\ or\ \frac{6}{21}$ c) $\frac{10}{15}\ or\ \frac{4}{6}$ d) $\frac{1}{3}\ or\ \frac{3}{100}$

24. Arrange these fractions into an increasing order of size:

a) $\frac{8}{3}, \frac{6}{4}\ and\ \frac{12}{5}$ b) $\frac{3}{10}, \frac{13}{20}\ and\ \frac{2}{3}$

25. Insert the symbol <, > or =:

a) $\frac{2}{15} \quad \frac{3}{10}$ b) $\frac{7}{30} \quad \frac{5}{12}$ c) $\frac{3}{20} \quad \frac{7}{12}$ d) $\frac{3}{20} \quad \frac{2}{15} \quad \frac{7}{10}$

3. Operations with fractions

Addition and subtraction of fractions

What is the sum of $\frac{1}{5}$ and $\frac{2}{5}$?

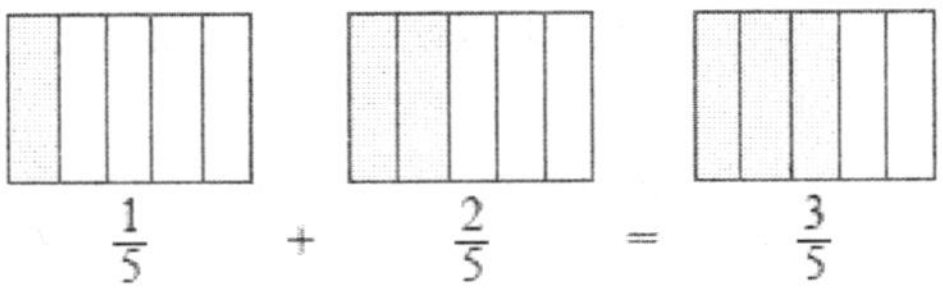

What is the difference of $\frac{3}{5}$ and $\frac{1}{5}$?

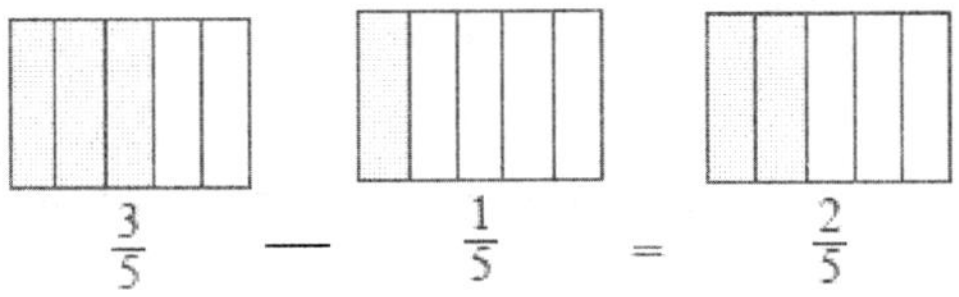

If the fractions have the same denominator the numerator of the sum is found by simply adding the numerators over the denominator. Their difference is the difference of the numerators over the denominator.

We do not add or subtract the denominators!!!

Reduce if necessary.

If the fractions have different denominators it is bit harder. Let´s see it with an example:

Example: Calculate $\frac{2}{5}+\frac{4}{7}$:

Solution:

Step 1: Find out the LCM(5, 7) = 35

Step 2: Transform both original fractions into two new fractions having 35 as denominator.

Remember you can multiply numerator and denominator by the same number:

$\frac{2}{5}=\frac{}{35} \Rightarrow$ To transform 5 into 35, we have multiplied by 7, so, we must also multiply the numerator by 7. So, as, $2 \cdot 7 = 14$, $\frac{2}{5}=\frac{14}{35}$

$\frac{4}{7}=\frac{}{35} \Rightarrow$ To transform 7 into 35, we have multiplied by 5, so, we must also multiply the numerator by 5. So, as, $4 \cdot 5 = 20$, $\frac{4}{7}=\frac{20}{35}$

Step 3: As new fractions have the same denominator, they can be added: $\frac{2}{5}+\frac{4}{7}=\frac{14}{35}+\frac{20}{35}=\boxed{\frac{34}{35}}$

In this case, the final fraction cannot be simplified.

Example: Calculate: $\frac{1}{4}-\frac{2}{3}+\frac{7}{5}$

Solution: LCM(4, 3, 5) = 60 $\rightarrow \frac{1}{4}-\frac{2}{3}+\frac{7}{5}=\frac{15}{60}-\frac{40}{60}+\frac{84}{60}=\boxed{\frac{59}{60}}$

Exercises

26. Evaluate the following, expressing your answers in the simplest form.

a) $\frac{2}{5}+\frac{1}{5}$ b) $\frac{3}{7}+\frac{2}{7}$ c) $\frac{3}{8}+\frac{2}{8}+\frac{5}{8}$ d) $\frac{3}{7}-\frac{2}{7}+\frac{4}{7}$

27. Evaluate the following, expressing your answers in the simplest form.

a) $\frac{1}{2}+\frac{3}{4}=$ b) $\frac{3}{5}+\frac{2}{10}+\frac{1}{15}=$ c) $\frac{7}{3}+\frac{5}{6}-\frac{1}{2}=$ d) $\frac{5}{2}+\frac{10}{3}-\frac{8}{3}-\frac{1}{3}=$

28. Joe painted 2/5 of a fence and Bill painted ½ of it. What fraction of the fence did the bys paint?

29. In a school $\frac{1}{3}$ of the children eat school dinners, $\frac{1}{2}$ bring packed lunches and the rest go home. What fraction of the children go home for lunch?

30. Sue bought a record with $\frac{1}{4}$ of her allowance. She spent another $\frac{1}{8}$ to see a movie. What part of her allowance did she spend?

31. A group of students went to a fast food restaurant. $\frac{2}{5}$ of them bought a beef burger and $\frac{1}{3}$ of them bought a chicken burger. The rest of them just bought drinks. What fraction of the group bought drinks?

Exercises

32. Last night John spent $\frac{3}{4}$ of an hour on English homework and $\frac{5}{6}$ of this time, on his Maths. If he started his homework at 20 o'clock, did he finish it in time to watch The Simpsons, which started at 21 o'clock? Explain your answer.

Sol.: Not, because $\frac{19}{12} > 1$.

Product of fractions

To ***multiply*** two or more ***fractions***, multiply the numerators and multiply the denominators, separately. Then simplify if necessary.

$$\frac{a}{b} \cdot \frac{c}{d} = \frac{a \cdot c}{b \cdot d}$$

Important: If possible, you must simplify the resulting fraction. You can do this at the end, but it will be more useful and comfortable for you if you simplify before multiplying. Let´s see it with an example:

Example: Calculate $\frac{6}{8} \cdot \frac{10}{15}$

Solution:

Simplifying at the end) $\frac{6}{8} \cdot \frac{10}{15} = \frac{6 \cdot 10}{8 \cdot 15} = \frac{60}{120} = \frac{6}{12} = \frac{6 \div 2}{12 \div 2} = \frac{3 \div 3}{6 \div 3} = \boxed{\frac{1}{2}}$

Simplifying before) $\frac{6}{8} \cdot \frac{10}{15} = \frac{6 \cdot 10}{8 \cdot 15} = \frac{(2 \cdot 3) \cdot (2 \cdot 5)}{(2 \cdot 2 \cdot 2) \cdot (3 \cdot 5)} = \frac{\not{2} \cdot \not{3} \cdot \not{2} \cdot \not{5}}{\not{2} \cdot 2 \cdot \not{2} \cdot \not{3} \cdot \not{5}} = \boxed{\frac{1}{2}}$

Note: Sometimes, it is possible to simplify even before: For example: Calculate $\frac{6}{10} \cdot \frac{10}{15}$

Solution: $\frac{6}{10} \cdot \frac{10}{15} = \frac{6 \cdot 10}{10 \cdot 15} = \frac{6}{15} = \frac{2 \cdot 3}{3 \cdot 5} = \boxed{\frac{2}{5}}$

Exercises

33. Multiply, giving your answers in the simplest form:

a) $42 \cdot \frac{5}{6}$ b) $\frac{5}{4} \cdot 37$ c) $\frac{5}{8} \cdot 18$ d) $2 \cdot \frac{3}{9} \cdot \frac{17}{5}$ e) $\frac{2}{8} \cdot 3 \cdot \frac{3}{7}$ f) $7 \cdot \frac{66}{21} \cdot \frac{17}{33}$

34. A math's teacher spends $\frac{400}{320}$ of his working time in school teaching his classes. If $\frac{80}{250}$ of this teaching time is spent with his first year classes. What fraction of his working time does he spend teaching the first year?

Calculating the fraction of a whole

Example: a) What is $\frac{2}{5}$ of 35? b) How much is the $\frac{5}{8}$ of 32 kg?

Solution: a) $\frac{1}{5}$ of $35 = 35 \div 5 = 7$ $\Rightarrow\Rightarrow\Rightarrow$ $\frac{2}{5}$ of $35 = 2 \times 7 = 14$.

b) $\frac{1}{8}$ of $32 = 32 \div 8 = 4$ $\Rightarrow\Rightarrow\Rightarrow$ $\frac{5}{8}$ of $32 = 5 \times 4 = 20$.

Exercises

35. Work out:

a) $\frac{1}{8}$ of 72 b) $\frac{4}{9}$ of 63 c) $\frac{5}{6}$ of 36 d) $\frac{3}{4}$ of 36

36. A chain store closed 2/15 of its 345 shops. How many shops were closed?

37. Mum bought 1200 g of grapes. John ate 2 fifths of them, Betty ate 1 quarter of them and dad ate the rest. What amount of grapes each of them eat?

38. Joan earns 1800 € a month. She spends 3/8 of her salary every month. She gives her parents 2/5 of the remainder and saves the rest. How much money does she save every month?

39. Mrs Holland spends 1/4 of her money in the market and 1/3 of the remainder in a shop. What fraction of her money is left?

40. Harban was given **£** 15 allowance each week. He spent 2/5 of it. What fraction did he save? How much did he save in pounds?

Inverse of a fraction

The ***inverse*** of a fraction $\frac{a}{b}$ is another fraction that multiplied by this one, their product is 1.

In an easier way, inverse of $\frac{a}{b}$ is $\boxed{\frac{b}{a}}$

Division of fractions

Quotient of two fractions is the product of the first one by the inverse of the second.

$$\frac{a}{b}:\frac{c}{d}=\frac{a}{b}\cdot\frac{d}{c}=\frac{a\cdot d}{b\cdot c}$$

Example: Calculate: $\frac{3}{8}:\frac{5}{3}$

Solution: $\frac{3}{8}:\frac{5}{3}=\frac{3}{8}\cdot\frac{3}{5}=\frac{3\cdot 3}{8\cdot 5}=\frac{9}{40}$

Important: If possible, you must simplify the resulting fraction. And you can do this at the end, as you already know, but you can also simplify before multiplying. Let´s see it with an example:

Example: Calculate $\frac{6}{8}:\frac{15}{10}$

Solution:

Simplifying at the end) $\frac{6}{8}:\frac{15}{10}=\frac{6}{8}\cdot\frac{10}{15}=\frac{6\cdot 10}{8\cdot 15}=\frac{60}{120}=\frac{6}{12}=\frac{6\div 2}{12\div 2}=\frac{3\div 3}{6\div 3}=\boxed{\frac{1}{2}}$

Simplifying before) $\frac{6}{8}:\frac{15}{10}=\frac{6}{8}\cdot\frac{10}{15}=\frac{6\cdot 10}{8\cdot 15}=\frac{(2\cdot 3)\cdot(2\cdot 5)}{(2\cdot 2\cdot 2)\cdot(3\cdot 5)}=\frac{2\cdot 3\cdot 2\cdot 5}{2\cdot 2\cdot 2\cdot 3\cdot 5}=\boxed{\frac{1}{2}}$

Note: Sometimes, it is possible to simplify even before:

For example: Calculate $\frac{6}{10}:\frac{15}{10}$

Solution: $\frac{6}{10}:\frac{15}{10}=\frac{6}{10}\cdot\frac{10}{15}=\frac{6\cdot 10}{10\cdot 15}=\frac{6}{15}=\frac{2\cdot 3}{3\cdot 5}=\boxed{\frac{2}{5}}$

Exercises

41. Work out, giving your answer in the simplest form:

a) $\frac{4}{5}:\frac{7}{3}$ b) $\frac{2}{3}:\frac{1}{3}$ c) $\frac{10}{9}:\frac{5}{3}$ d) $\frac{5}{2}:\frac{1}{2}$ e) $\frac{7}{3}:\frac{2}{3}$ f) $\frac{8}{3}:\frac{6}{4}$

42. Calculate, giving your answer in the simplest form:

a) $\frac{3}{2}:\frac{4}{9}$ b) $\frac{1}{5}:\frac{5}{19}$ c) $\frac{3}{14}:\frac{4}{7}$ d) $\frac{15}{3}:\frac{5}{6}$ e) $\left(\frac{5}{3}:\frac{5}{8}\right):\frac{7}{4}$ f) $2:\left(\frac{9}{7}:\frac{13}{3}\right)$

43. In a factory a metal rod 49 meter long is divided into $\frac{7}{5}$ meter pieces. How many pieces can be made from each rod?

44. A six tons load of soil container is divided in $\frac{3}{140}$ tons bags. How many bags can be filled?

4. Fractions and decimal numbers

Decimal expression of rational numbers

As a fraction is the indicated quotient of two numbers, we can calculate these quotients only by dividing numerator and denominator.

Let´s find the decimal expressions of fractions $\frac{7}{4}$, $\frac{7}{3}$ and $\frac{11}{6}$.

```
7   |4              7   |3              11   |6
30   1'75           10   2'333           50   1'8333
 20                  10                   20
  0                   10                   20
                                            20
```

So, these decimal expressions are:

$$\frac{7}{4}=1.75 \qquad \frac{7}{3}=2.333........ \qquad \frac{11}{6}=1.8333......$$

As we have seen in above examples, when dividing numerator by denominator there are these situations:

- Division finishes because it is exact, as happens in the first fraction. In this case, we obtain a **limited** decimal expression (it has a limited number of decimal digits). Its decimal expression is said to be an ***exact or terminating decimal number***.

- Division does not finish because it is not exact, as happens in second and third fractions. In these cases, we obtain a **non-limited** decimal expression (its decimal digits do not finish), and they have a repeating part. Their decimal expressions are named ***recurring*** or ***repeating decimal numbers*** Second one is named ***pure recurring decimal number***, and third one is a mixed recurring decimal number.
Repeating part is named ***period*** and it is indicated with an arch on it. For example:

$$2.333........ = 2.\overline{3} \qquad\qquad 1.8333........ = 1.8\overline{3}$$

- Division does not finish, and, furthermore, there is not a repeating decimal part.

So, decimal numbers can be classified into:

– **Exact or terminating** decimal: It finishes, so you can write down all its digits.
For example: 0,125

– **Recurring or repeating** decimal: It does not finish, it goes on forever, but some of the digits are repeated over and over again. For example:

$5.6666...... = 5.\overline{6}$ $\qquad$ $14.353535....... = 14.\overline{35}$ $\qquad$ $87.42222...... = 87.4\overline{2}$

We can distinguish:

- **Pure recurring decimals**: Recurring decimals in which period starts just after the decimal point. For example: $5.6666...... = 5.\overline{6}$ $\qquad$ $14.353535....... = 14.\overline{35}$
- **Mixed recurring decimals:** Recurring decimals in which period does not start just after the decimal point. For example: $87.42222...... = 87.4\overline{2}$

– **Irrational** numbers: Decimals that cannot be written as fractions. They do not finish and they do not have any repeating part. For example: $\sqrt{2} = 1.4142135....$ $\pi = 3.141592654.....$ $e = 2.718.......$

Exercises

45. Express in its decimal form the following fractions and indicate their type of decimal number: a) $\frac{3}{4}$ b) $\frac{4}{3}$ c) $\frac{3}{5}$ d) $\frac{5}{6}$ e) $\frac{7}{15}$ f) $\frac{5}{7}$

46. Write 12 decimal digits for each of the following recurring decimals:

a) $3.\overline{246}$ b) $4.\overline{6}$ c) $2.43\overline{28}$ d) $3.2\overline{478}$

47. Express the following fractions in their decimal form. Observe prime factors of their denominators. Could you obtain any conclusion?

a) $\frac{3}{4}$ b) $\frac{7}{5}$ c) $\frac{23}{20}$ d) $\frac{13}{25}$ e) $\frac{23}{15}$

48. We are going to help you with previous exercise. Complete the following statement:

- "*Decimal fraction:* Fractions having denominator whose only prime factors are ___ and ___, have as decimal form, an exact decimal".

- "So, we can assure that if denominator of a fraction has a prime factor different to ___ or ___, its decimal form will be a recurring or an irrational decimal".

Rational expression of decimal numbers

Exact and recurring decimals can be expressed as a fraction.

In following example we show the process to find out these fractions.

a) Exact decimal: Express as a fraction $x = 2.34$

Solution: We must build a fraction whose

- Numerator: Decimal number without decimal point.
- Denominator: A "1" followed by as many "0" as decimal digits decimal number has.

In our case: $2.34 = \frac{234}{100}$ If possible, we must simplify this fraction to give it in its simplest form: $2.34 = \frac{224}{100} = \frac{112}{50} = \boxed{\frac{56}{25}}$

b) Recurring decimal number: Express as a fraction $x = 14.\overline{35} = 14.3535355.........$

Solution: We must apply the following process, consisting on multiplying it by 1, 10, 100, 1000, …., till finding two decimals having *identical decimal parts*, so that they can be subtracted, to cancel their decimal parts and converting them into integers.

$$100 \cdot x = 1435.353535....$$
$$10 \cdot x = 143.535353.....$$
$$x = 14.353535........$$

As you can see, x and 100·x have the same decimal part. So, we subtract them:

$$100 \cdot x = 1435.353535....$$
$$\require{cancel}\cancel{10 \cdot x = 143.535353.....}$$
$$x = 14.353535........$$

$$99 \cdot x = 1421 \Rightarrow x = \frac{1421}{99}$$

Example: Express as a fraction: a) $8.\overline{5}$ b) $25.4\overline{52}$

Solution:

a)

$$10 \cdot x = 85.555.......$$
$$x = 8.5555......$$

As you can see, both decimals have the same decimal part. So, we subtract them:

$$10 \cdot x = 85.555.......$$
$$x = 8.5555......$$

$$9 \cdot x = 77 \Rightarrow x = \frac{77}{9}$$

b)

$$1000 \cdot x = 25452.5252......$$
$$100 \cdot x = 2545.25252......$$
$$10 \cdot x = 254.525252....$$
$$x = 25.4525252....$$

As you can see, 1000·x and 10·x have the same decimal part. So, we subtract them:

$$1000 \cdot x = 25452.5252......$$
$$\cancel{100 \cdot x = 2545.25252......}$$
$$\cancel{10 \cdot x = 254.525252....}$$
$$x = 25.4525252....$$

$$990 \cdot x = 25198 \Rightarrow x = \frac{25198}{990}$$

Exercises

49. Express as a fraction:

a) 23.45　　b) 5.101

50. Express as a fraction:

a) $24.\overline{3}$　　b) $0.\overline{678}$　　c) $32.\overline{76}$　　d) $1.\overline{876}$　　e) $26.\overline{7}$

51. Express as a fraction:

a) $2.7\overline{5}$　　b) $34.6\overline{41}$　　c) $2.09\overline{5}$　　d) $0.2\overline{3}$　　e) $5.1\overline{204}$

52. Express as a fraction:

a) $1.0\overline{4}$　　b) $7.\overline{123}$　　c) $8.17\overline{4}$　　d) $1.7\overline{6}$　　e) $2.5\overline{19}$

Rounding decimal numbers

In some situations it is more reasonable working with *approximations* than with exact values. Read the following examples:

- ***Zero number discovery occurred by year 400.***
- ***By X century, Córdoba had a population of 500.000 inhabitants.***
- ***By year 2050, only 41% of world population will live in country zones.***

As you can see, we are given an *approximation* of an unknown quantity. This kind of approximations are named ***estimations***.

Now, look at these examples:

- If a TV news program wants to inform that, in Spain, incomes derived from tourism have been 235.467.342 €, they will do like: ***«In Spain, incomes obtained by tourism have been 235 M€».***

- If quotient 13/6 = 2.1666… we eliminate all decimal digits from second one, and we obtain the decimal 2.16. This number is not exactly 13/6, but its value is only lightly less than the value of 13/6.

Even knowing we are making a little ***mistake,*** we can use it as a substitute in determined calculations.

If we eliminate decimal digits from third one, we obtain the decimal 2.166, more approximated to exact value.

Approximation degree depends on the desired objective. For example, if distance between two towns is 101 km, and we say distance is 100 km, error of 1 km is acceptable; but in precision pieces design, it is required a lower error, from 0.01 mm to 0.001 mm.

One of the methods to approximate a number is ***rounding.*** To round a number, we must pay attention to the first digit we want to cancel:

To round a decimal number to a given place value

1. Draw an arrow under the given place value. (After enough practice, you will be able to round mentally and will not need the arrow.)

2. If the digit to the right of the arrow is 5, 6, 7, 8, or 9, add 1 to the digit above the arrow; that is, round to the larger number.

3. If the digit to the right of the arrow is 0, 1, 2, 3, or 4, keep the digit above the arrow; that is, round to the smaller number.

4. Write whatever zeros are necessary after the arrow so that the number above the arrow has the same place value as the original.

Example:

a) Round 0.7539 to the nearest hundredth.

$0.7539 \approx 0.75$ ↑ — The digit to the right of the round-off place is 3, so round down.

b) Round 7843.9 to the nearest thousand.

$7843.9 \approx 8000$ ↑ — Three zeros must be written after the 8 to keep it in the thousands place.

c) Round 537.7 to the nearest unit.

$537.7 \approx 538$ ↑ — The number to the right of the round-off place is 7, so we round up by adding 537 + 1 + 538.

Example: Round

a) 3.25 to one decimal (tenths) → 3.3.

b) 0.4444..... to two decimals (hundredths) → 0.44.

c) 1.77777.........to two decimals (hundredths) → 1.78.

Exercises

53. Round off the numbers:

a) 2.36 to the nearest tenth

b) 6.3757 to the nearest hundredth

c) 0.8903 to the nearest thousandth

d) 17.17 to the nearest unit

e) 176.705 to the nearest hundred

f) 67,756 to the nearest thousand

54. Copy in your notebook the following table and round to tenths, hundredths and thousandths:

		Tenths	Hundredths	Thousandths
Rounding	3.7125			
	21.2651			
	0.5352			
	203.4721			
	7.0181			

Exercises

55. Calculate deviation or mistake we make when rounding 19/7 to tenths.

56. a) Calculate the following addition rounding previously the quantities to hundredths, and calculate the mistake you are making: 4.685 + 27.0392 + 0.962.

b) Calculate the subtraction rounding previously the quantities to tenths, and calculate the mistake you are making: 28784.62 – 1475.983.

57. Write the decimal point in the results of the following operation, without calculating them:

a) 23.64 + 17.125 = 40765 d) 15.92 – 7.874 = 8046 g) 13.9 · 0.5 · 2.7 = 171665

b) 0.42 + 8.5 = 892 e) 3.28 · 4.15 = 13612 h) 123.25 : 2.9 = 425

c) 4.025 – 3.15 = 875 f) 16.5 · 2.08 = 3432 i) 3.9285 : 0.45 = 873

Review exercises

1. Use integers to express the following quantities:

a) The depth of a vulcano, 600 m.

b) The second floor of an underground parking.

c) An income in my bank account, with value of 4000 €.

2. Calculate:

a) (+ 5) + (+ 2) b) (+ 3) + (- 2) c) (+ 5) + (- 9) d) (- 2) + (+ 12)

e) (- 7) + (+ 6) f) (- 10) + (- 5) g) (+ 3) – (+ 5) h) (+ 15) – (+ 5)

i) (+ 7) – (- 3) j) (- 5) – (+ 5) k) (- 13) – (- 3)

3. Work out:

a) $-18-(-9)$ b) $15-(-10+18)$ c) $-12-(-12-4)$

d) $-(-12)-(12+7)$ e) $-7+32$ f) $-12+8+(-10)+4$

g) $-(-8)+12+(-5)+7$ h) $-15+30+(-19)+20$ i) $-15+7+(-8)+9$

j) $15+(-10)+5+(-10)$ k) $-12+8+(-7)+(-2)$ l) $-13+15+(-10)+9$

m) $-20-(-8+4-5)$ n) $12-(-8+10)$ ñ) $-(-8)-(4-7-9)$

o) $-15-(-2+9)$

4. Calculate the following products and quotients:

a) $(+5)\cdot(+3)$ b) $(+8)\cdot(-4)$ c) $(-6)\cdot(+2)$ d) $(-5)\cdot(-4)$ e) $(+4)\cdot(-2)$

f) $(-1)\cdot(-1)\cdot(-1)$ g) $(-2)\cdot(+3)\cdot(-6)$ h) $(+5)\cdot(-2)\cdot(+3)$ i) $(+40):(-20)$ j) $(-15):(+3)$

k) $(-21):(-3)$ l) $(+10):(+5)$

5. Calculate:

a) $-65+13$ b) $6-(-23)$ c) $-2+(-3)$ d) $(-12)\times(-3)\times 5$

e) $45:(-5)$ f) $3-2+9-7$ g) $3-2\times 5+7\times 4$ h) $7+(4-5)-(-89)$

i) $4-3+7-2$ j) $(-3)\times(-2)+7$ k) $3+4\times 5$ l) $-8:4-1$

m) $(-1)\times 7+(2-5)$ n) $4:2-(3+1)$ ñ) $13-(4+8)-3\times 54$

o) $[21:(7\times 3)]+4\times(5-1)$ p) $7-7-[(2\times 3):(3\times 2)]$

q) $-3-2\times[-9\times(5-4)-(-6)]$ r) $-3+3\times(5-(-4))$ s) $11\times 5-6\times 11$

6. Calculate:

a) $9+5\cdot 3-12$ b) $120:2-5\cdot(8+2-10)$ c) $6\cdot 7-35:7+2\cdot 8$

d) $46:2-10\cdot(16-3\cdot 5)$ e) $(12+8):2+2\cdot(7+3)-2\cdot 12$ f) $7\cdot 3-49:7+5\cdot(8-5)$

g) $10-10:2+15:3+4\cdot 4$ h) $(4+8-3+5)\cdot 5+2$ i) $(6+8-4):2+18:(5+4)$

j) $350:5-15\cdot 4+65:5$ k) $125+68\cdot 2-120:4+100$

7. Find the $\frac{2}{3}$ of 12 and, later, the $\frac{4}{6}$. Why do you obtain the same value?

8. In a bag, 2 balls of each 5 are blue. If there are 12 blue balls, how many balls are there in the bag?

9. Calculate:

a) The $\frac{4}{7}$ of 21 b) The $\frac{3}{5}$ of 70 c) The $\frac{7}{3}$ of 81 d) The $\frac{2}{3}$ of $\frac{5}{6}$

10. Find the simplest form of the following fractions:

a) $\frac{40}{45}$ b) $\frac{60}{72}$ c) $\frac{200}{240}$ d) $\frac{780}{2964}$

11. Find the simplest form of the following fractions:

a) $\frac{45}{27}$ b) $\frac{24}{60}$ c) $\frac{64}{72}$ d) $\frac{4+6}{4+8}$

12. Compare these fractions:

a) $\frac{5}{8}$ *and* $\frac{3}{7}$ b) $\frac{7}{6}$ *and* $\frac{4}{9}$ c) $\frac{1}{2}, \frac{3}{4}$ *and* $\frac{5}{6}$ d) $\frac{3}{14}, \frac{3}{35}$ *and* $\frac{1}{7}$

13. In a schotl, 1/3 of the students study English and the 33% French. What is the most chosen language?

14. A car advances 50 km in three quarters of an hour, and another one, 36 km in 27 minutes. Which one does advance more quicly?

15. A student says: «During a day, I spend 1/3 of time sleeping and 3/8 at school. Classes take the 2/3 of the time I spend at school». a) What fraction of my day do classes take? b) How many hours?

16. Work out:

a) $\frac{4}{5}+\frac{1}{6}-\frac{2}{15}$ b) $2+\frac{5}{6}+\frac{1}{3}$ c) $\frac{-3}{10}+\frac{5}{6}+\frac{4}{3}$ d) $\frac{-5}{12}+\frac{2}{9}+\frac{-7}{15}$ e) $\left(\frac{3}{4}-\frac{1}{5}\right)-\left(\frac{-7}{2}+\frac{5}{2}\right)$

17. Work out:

a) $\frac{2}{7}\cdot\frac{3}{5}$ b) $\frac{2}{3}\cdot\frac{5}{7}\cdot\frac{6}{9}$ c) $\frac{2}{7}:\frac{3}{5}$ d) $\left(\frac{2}{7}:\frac{4}{5}\right)\cdot\frac{4}{6}$ e) $\frac{15}{2}:(-5)$ f) $\left(\frac{3}{4}\cdot\frac{1}{5}\right)+\left(\frac{-7}{2}:\frac{5}{2}\right)$

18. Work out by two methods, directly and extracting a common factor:

a) $\frac{2}{3}\cdot\frac{1}{5}+\frac{2}{3}\cdot\frac{5}{6}$ b) $\frac{3}{4}\cdot\frac{5}{6}+\frac{5}{6}\cdot\frac{1}{9}$ c) $\frac{-3}{11}\cdot\frac{1}{4}-\frac{6}{5}\cdot\frac{-3}{11}$

19. Work out the following expressions:

a) $3-4\left[\frac{1}{3}-\frac{1}{2}\left(\frac{1}{4}-\frac{1}{5}\right)+3:\left(\frac{1}{3}:\frac{1}{2}\right)\right]$

b) $(3-4)\left[\left(\frac{1}{3}-\frac{1}{2}\right)\left(\frac{1}{4}-\frac{1}{5}\right)+\left(3:\frac{1}{3}\right):\frac{1}{2}\right]$

c) $\left(\frac{1}{3}+\frac{1}{2}\right)\left(\frac{1}{2}-\frac{1}{4}\right)+5-3\left(4:\frac{3}{5}+1\right)$

d) $\left[\frac{1}{3}+\frac{1}{2}\left(\frac{1}{2}-\frac{1}{4}\right)+5\right]-3\left[4:\left(\frac{3}{5}+1\right)\right]$

e) $\left[\frac{3}{8}\left(\frac{5}{3}-\frac{1}{2}\right)-\frac{4}{11}\left(\frac{3}{4}-\frac{1}{5}\right)\right]:\left[\frac{5}{9}-\left(\frac{-3}{4}+\frac{1}{2}\right)+\frac{10}{3}\left(\frac{1}{2}-\frac{3}{5}\right)\right]$

f) $\left(\frac{3}{5}:\frac{2}{3}-\frac{4}{5}\cdot\frac{4}{3}+\frac{1}{3}-\frac{3}{4}:\frac{3}{7}\right)\left(\frac{2}{3}+\frac{-7}{2}-\frac{5}{6}+\frac{1}{4}\right):\left(\frac{-4}{3}+\frac{2}{3}-\frac{1}{6}\right)$

20. A bottle has 3/4 liter of orange juice, another one has 3/5 and a third one has 5/6. a) What is the total quantity of orange? b) How much juice has the first bottle more than the second one?

21. Temperatures in scales *Celsius (C)* and *Fahrenheit (F)* are related by the expression $°F = \frac{9}{5} \cdot °C + 32$.

a) If temperature of an ill person is 40 °C, what would his temperature be in a Fahrenheit termometer?

b) Boiling point of water is 100 °C. What is this temperature in Fahrenheit scale?

c) Find out the vaue of F for C = 20° and C = 50°.

22. A family spends 1/16 of their month money in renting their house, 1/64 in telephone and electricity and 1/8 in transport and cloths.

a) What portion of month money is spent in previous items?

b) What portion do they have for food and saving?

c) How do they distribute their spenses if their month incomes are 1082 €?

d) How much money do they have for food and saving?

23. Express in its decimal form the following fractions and indicate their type of decimal number:

a) $\frac{3}{4}$ b) $\frac{4}{3}$ c) $\frac{3}{5}$ d) $\frac{5}{6}$ e) $\frac{7}{15}$

24. Write 12 decimal digits for each of the following recurring decimals:

a) $3.\overline{246}$ b) $4.\overline{6}$ c) $2.43\overline{28}$ d) $3.2\overline{478}$

25. Express as a fraction:

a) $23.\overline{45}$ b) $76.8\overline{1}$ c) 398.12 d) 5.6 e) $28.\overline{7}$

f) $261.1\overline{23}$ g) $0.0\overline{4}$ h) 23.05

26. Express as a fraction:

a) 2.41 b) 0.75 c) $3.\overline{5}$ d) $1.0\overline{3}$

27. Express as a fraction and check that the result of these operations is an integer:

a) 2.333… + 4.666…..   b) 6.171717… + 3.828282….

28. *CURIOSITY.* Express the following decimals as fraction:

$0.\overline{9}$ $5.\overline{9}$ $0.0\overline{9}$ $12.5\overline{9}$

What conclusion do you obtain?

29. Calculate the decimal expressions for the following fractions and round it to the thousandths:

a) 3/7   b) 4/3   c) 17/3   d) 2/7

30. Round the following numbers:

a) 45.00768 to the hundredths   b) 0.987125 to the thousandths

c) 5.6 to the units

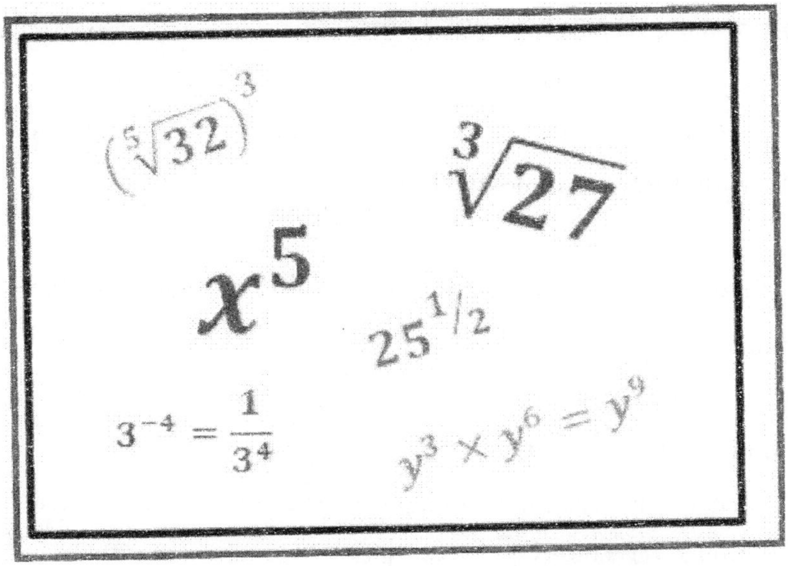

Unit 2.- Powers and roots

1. Powers

A **Power** is a number obtained by multiplying a number by itself a certain number of times

$$a^n = \underbrace{a \cdot a \cdot a \cdot a \cdot \ldots\ldots a}_{n\ times} \quad where \begin{cases} a: base \\ b: exponent \end{cases}$$

For example, $6^5 = 6 \cdot 6 \cdot 6 \cdot 6 \cdot 6$, and can be read as:

- Six to the fifth power
- Six to the power of five
- Six powered to five.

The most common is **six to the power of five.**

The number that is successively multiplied by itself is called the **base**. A small raised number called the **exponent** follows the base and indicates the number of times the base is to be multiplied.

And, remember: $\boxed{a^1 = a}$ and $\boxed{a^0 = 1}$, for any natural number *a*.

Special cases: Squares and cubes

3^2 is read as three to the second power, three to the power of two, three to the square power, but the most common is **three squared.**

5^3 is read as five to the third power, five to the power of three, but the most common is **five cubed.**

Operations with powers

To manipulate expressions with powers we use some rules that are called **laws of powers.**

Laws	Examples
$a^m \cdot a^n = a^{(m+n)}$	$a^4 \cdot a^3 = (a \cdot a \cdot a \cdot a) \cdot (a \cdot a \cdot a) = a \cdot a \cdot a \cdot a \cdot a \cdot a \cdot a = a^7$
$\frac{a^m}{a^n} = a^{(m-n)}$	$\frac{a^8}{a^3} = \frac{a \cdot a \cdot a \cdot a \cdot a \cdot \not a \cdot \not a \cdot \not a}{\not a \cdot \not a \cdot \not a} = a \cdot a \cdot a \cdot a \cdot a = a^5$
$(a^m)^n = a^{m \cdot n}$	$(a^4)^2 = a^4 \cdot a^4 = a \cdot a \cdot a \cdot a \cdot a \cdot a \cdot a \cdot a = a^8$
$(a \cdot b)^m = a^m \cdot b^m$	$(a \cdot b)^3 = (a \cdot b) \cdot (a \cdot b) \cdot (a \cdot b) = a \cdot b \cdot a \cdot b \cdot a \cdot b = (a \cdot a \cdot a) \cdot (b \cdot b \cdot b) = a^3 \cdot b^3$
$\left(\frac{a}{b}\right)^m = \frac{a^m}{b^m}$	$\left(\frac{a}{b}\right)^3 = \frac{a}{b} \cdot \frac{a}{b} \cdot \frac{a}{b} = \frac{a \cdot a \cdot a}{b \cdot b \cdot b} = \frac{a^3}{b^3}$

Exercises

1. Calculate the value of the following powers:

a) 3^4 b) 4^5 c) 6^2 d) $(-4)^3$ e) 5^6 f) $(-5)^5$ g) 2^{10} h) $(-3)^6$

2. Write as an only power:

a) $2^3 \cdot 2^4$ b) $3^5 \cdot 3^3 \cdot 3^4$ c) $(-7)^2 \cdot (-7)^5$ d) $4^5 : 4^2$ e) $7^6 : 7^3$

f) $x^9 \cdot x^4$ g) $(5^2)^3$ h) $((-2)^2)^3$ i) $(-2)^9 : (-2)^5$ j) $(-9)^7 : (-9)^4$

3. Write as an only power:

a) $5^4 \cdot 3^4$ b) $2^2 \cdot 4^2$ c) $3^4 \cdot 4^4 \cdot 5^4$ d) $x^n \cdot y^n$ e) $(-5)^4 \cdot (-3)^4$

f) $8^3 : 4^3$ g) $9^5 : 3^5$ h) $x^n : y^n$ i) $(-10)^4 : 2^4$ j) $(-6)^3 : (-3)^3$

Exercises

4. Substitute the symbol (Δ) by a number:

a) $2^{\Delta} \cdot 2^{11} = 2^{20}$ b) $(-4)^6 : (-4)^{\Delta} = (-4)^3$ c) $(-6)^3 \times (-6)^5 = (-6)^{\Delta}$

d) $[(-1)7]^{\Delta} = (-1)^{21}$ e) $(3^{11})^2 = 3^{\Delta}$

5. Write as an only power:

a) $4^3 \cdot 4^4 \cdot 4$ b) $(5^6)^3$ c) $\dfrac{7^6}{7^4}$ d) $\dfrac{15^3}{3^3}$

e) $2^{10} \cdot 5^{10}$ f) $\dfrac{12^5}{3^5 \cdot 4^5}$ g) $(a^6 \cdot a^3)^2 : (a^2 \cdot a^4)^3$ h) $(6^2)^3 \cdot 3^5 \cdot (2^7 : 2^2)$

6. Calculate by using the laws of powers:

a) $2^3 \cdot 5^4$ b) $(6^5 : 2^4) : 3^5$ c) $\left(\dfrac{2}{3}\right)^6 \cdot \left(\dfrac{3}{4}\right)^3$ d) $2^8 \cdot \left(\dfrac{5}{2}\right)^4$

e) $\dfrac{20^6}{2^6}$ f) $\dfrac{20^6}{2^5}$ g) $(3^3)^2 : 3^5$ h) $(2^5)^3 \cdot [(5^3)^4 : 2^3]$

Powers having an integer exponent

Till now, we have only studied powers having exponents > 1. This might make you wonder if powers like 5^1, 8^0 or 7^{-2} exist. Answer is affirmative. They exist and their values are:

$$a^1 = a \qquad a^0 = 1 \qquad a^{-n} = \frac{1}{a^n} \qquad \left(\frac{a}{b}\right)^{-n} = \left(\frac{b}{a}\right)^n$$

In the next lines, we are going to demonstrate these values:

Exponent = 1

Choose two numbers whose difference is 1. For example, 4 and 3. ➡ $a^1 = a^{4-3} = \frac{a^4}{a^3} = \frac{a \cdot \cancel{a} \cdot \cancel{a} \cdot \cancel{a}}{\cancel{a} \cdot \cancel{a} \cdot \cancel{a}} = a$ ➡ $\boxed{a^1 = a}$

Exponent = 0

Choose two numbers whose difference is 0. For example, 3 and 3. ➡ $a^0 = a^{3-3} = \frac{a^3}{a^3} = \frac{\cancel{a} \cdot \cancel{a} \cdot \cancel{a}}{\cancel{a} \cdot \cancel{a} \cdot \cancel{a}} = 1$ ➡ $\boxed{a^0 = 1}$

Negative exponent

Choose two numbers whose difference is negative. For example 3 and 5 (3 – 5 = -2). ➡ $a^{-2} = a^{3-5} = \frac{a^3}{a^5} = \frac{\cancel{a} \cdot \cancel{a} \cdot \cancel{a}}{a \cdot a \cdot \cancel{a} \cdot \cancel{a} \cdot \cancel{a}} = \frac{1}{a^2}$ ➡ $\boxed{a^{-2} = \frac{1}{a^2}}$

And also: $a^{-1} = \left(\frac{a}{1}\right)^{-1} = \left(\frac{1}{a}\right)^{1} = \frac{1}{a}$

Example: Calculate: a) 7^{-2} b) $5^{-3} \cdot 5^3$ c) $2^{-2} \cdot 3^{-2}$ d) $(2^3)^{-2}$ e) $(-6)^{-3}$ f) $8^2 : 8^{-2}$ g) $12^{-3} : 4^{-3}$

Solution:

a) $7^{-2} = \frac{1}{7^2} = \frac{1}{49}$

b) $5^{-3} \cdot 5^3 = 5^{-3+3} = 5^0 = 1$

c) $2^{-2} \cdot 3^{-2} = (2 \cdot 3)^{-2} = 6^{-2} = \frac{1}{6^2} = \frac{1}{36}$

d) $(2^3)^{-2} = 2^{-6} = \frac{1}{2^6} = \frac{1}{64}$

e) $(-6)^{-3} = \frac{1}{(-6)^3} = \frac{1}{-216} = -\frac{1}{216}$

f) $8^2 \cdot 8^{-2} = 8^{2-2} = 8^0 = 1$

g) $12^{-3} : 4^{-3} = (12/4)^{-3} = 3^{-3} = \frac{1}{3^3} = \frac{1}{27}$

Exercises

7. Calculate the following powers:

a) $5^6 \cdot 5^{-3}$ b) $(-4)^{-6} : (-4)^{-2}$ c) $[(-3)^2]^{-5}$ d) $4^3 \cdot 4^{-3}$

e) $(-75)^3 : 5^3$ f) $(6^{-2})^{-5}$ g) $8^6 \times (-4)^6$ h) $(-3)^5 : (-3)^4$

Exercises

8. Calculate the value of the following expressions:

a) $2^2 - 4^2 : 8 + 30$ b) $2 \cdot 3^2 - 5^2 : 5 + 5^3$ c) $3^{-1} \cdot 3 - 3^0 + 1 - 25^1$

d) $32 : 2 - 1 - 32 : 2\text{-}1$ e) $\frac{3^2}{3} - \frac{2^{-1}}{3} + \frac{1}{3^{-1}}$ f) $\frac{2^2}{3} - 1 - \frac{3^2}{2^{-1}}$

9. Give the following fractions in their simplest form:

a) $\frac{(-3)^2 \cdot (-5) \cdot (-2)^3}{-8}$ b) $\frac{(-3)^2 \cdot (-2)^3 \cdot 2^2}{1}$ c) $\frac{(-15)^3 \cdot (-3)^{-1}}{(-3)^5 \cdot 5^2}$ d) $\frac{(-6)^3 \cdot 9^2 \cdot (-2)^6}{(-12)^5}$

10. Write as a simplified fraction:

a) $\frac{3^4}{3^5}$ b) 5^{-1} c) a^{-6} d) $x^{-1}y^{-2}$ e) $\frac{x^3 y^4}{x^2 y^6}$ f) $(3xy^2)^{-2}$ g) $5 \cdot 3^{-1} \cdot xy^{-2}$

11. Write as an only rational number:

a) $\left(\frac{1}{5}\right)^2$ b) $\left(\frac{1}{5}\right)^{-2}$ c) $\left(\frac{-1}{5}\right)^{-2}$ d) $\left(\frac{3}{4}\right)^{-2}$ e) $\left(\frac{1}{5} \cdot \frac{1}{2}\right)^{-6}$

f) $\left(\frac{1}{2}\right)^6 \cdot \left(\frac{1}{5}\right)^6$ g) $\left(\frac{2}{3}\right)^3 \cdot \left(\frac{2}{3}\right)^2$ h) $\left(\frac{17}{45}\right)^0$ i) $\left[\left(\frac{1}{3}\right)^{-3}\right]^2$

Scientific notation

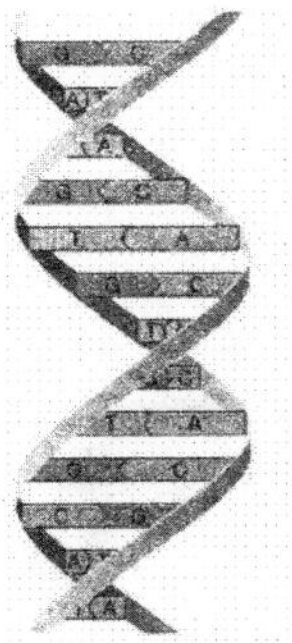

When we observe Nature out of quotidian life, for example going into microscopic world or to other planets and galaxies, we are finding quantities that are not comfortable to use. Sometimes because they are very little (size of a cell = 0.0000000018 cm) and sometimes because they are too large (light speed = 300 000 000 m/s).

These quantities, containing long chains of zeros, to the right or to the left of significant digits, make them uncomfortable to use.

In order to make them more comfortable to handle, it is used ***scientific notation***, which consists on transforming numbers with long chains, into shorter expressions. In the following examples, we are showing how scientific notation works.

Parts of a number in scientific notation, $N = a.bcc... \times 10^k$:

· An integer part, with an only digit (Not zero).

· A decimal part.

· A power with base 10 and integer exponent. (***k*** is an integer, positive or negative).

Example: Express 322400000000000 in scientific notation.	
<u>Step 1:</u> Write significant digits, with a decimal point after the first of them. ⇨	3.224
<u>Step 2:</u> Count how many positions the decimal point has been moved. This is going to be the exponent. ⇨	322400000000000.0000……. 3,2240000000000000000000 We have moved decimal point 14 positions. So, $3.224 \cdot 10^{14}$
<u>Paso 3º:</u> <u>Sign of the exponent.</u> There is a very easy rule: - If initial quantity was higher than new one with decimal point, exponent is positive; ⇨ - If initial quantity was lower than new one with decimal point, the exponent will be negative.	In this case, as 322400000000000 > 3,224, exponent stays positive: $3.224 \cdot 10^{14}$

This transformation can also be done by using the definition of power:

Example: Express in scientific notation these quantities: a) 529000 b) 0.00025

<u>Solution:</u>

a) $529000 = 5.29 \cdot 100000 = 5.29 \cdot 10^5$

b) $0.00025 = \frac{2.5}{100000} = 2.5 \cdot \frac{1}{100000} = 2.5 \cdot 10^{-5}$

Exercises

12. Write in scientific notation:

$6,57\cdot10^{2}$	$5,8\cdot10^{-4}$	$1,258\cdot10^{6}$	$2,1\cdot10^{-3}$	$3,21\cdot10^{5}$
a) 657	b) 0.00058	c) 1258000	d) 0.0021	e) 321000
f) 0.000012	g) 0.0012	h) 7800	i) 9757000	j) 0.00023
$1,2\cdot10^{-5}$	$1,2\cdot10^{-3}$	$7,8\cdot10^{3}$	$9,757\cdot10^{6}$	$2,3\cdot10^{-4}$

Product and quotient in scientific notation

To ***multiply*** or ***divide*** two quantities given in scientific notation, we multiply or divide, separately, numeric parts and powers of 10. You will have to use the laws of powers.

Example: Calculate: a) $2.7\cdot10^{-5}\cdot 3.2\cdot10^{8}$ b) $\dfrac{4.62\cdot10^{7}}{2\cdot10^{-4}}$ c) $8.31\cdot10^{-5}\cdot 5.2\cdot10^{9}$

Solution:

a) $2.7\cdot10^{-5}\cdot 3.2\cdot10^{8} = \left| \begin{matrix} 2.7\cdot3.2 = 8.64 \\ 10^{-5}\cdot10^{8} = 10^{-5+8} = 10^{3} \end{matrix} \right| = \boxed{8.64\cdot10^{3}}.$

b) $\dfrac{4.62\cdot10^{7}}{2\cdot10^{-4}} = \left| \begin{matrix} \dfrac{4.62}{2} = 2.31 \\ 10^{7}:10^{-4} = 10^{7-(-4)} = 10^{11} \end{matrix} \right| = \boxed{2.31\cdot10^{11}}.$

c) $8.31\cdot10^{-5}\cdot 5.2\cdot10^{9} = \left| \begin{matrix} 8.31\cdot5.2 = 43.212 \\ 10^{-5}\cdot10^{9} = 10^{-5+9} = 10^{4} \end{matrix} \right| = 43.212\cdot10^{4}$, but in scientific notation, we can write only one significant digit before decimal point.

So, we will write $4.3212\cdot10\cdot10^{4} = \boxed{4.3212\cdot10^{5}}$.

Exercises

13. Work out:

a) $4,1\cdot10^{3}$ x $2\cdot10^{5}$ $8,2\cdot10^{8}$

b) $2,8\cdot10^{-4} : 2\cdot10^{3}$ $1,4\cdot10^{-7}$

c) $3,2\cdot10^{5} : 2\cdot10^{-2}$ $1,6\cdot10^{7}$

Addition and subtraction of numbers in scientific notation

When ***adding*** or ***subtracting*** numbers in scientific notation, we need all the powers of 10 to have the ***same exponents***. In this case, we will add or subtract the numerical part, leaving the same power of 10.

* If they do not have the same exponent, previously, we must reduce to <u>*common exponent*</u>.

Example: Calculate: a) $3.75 \cdot 10^8 + 2.11 \cdot 10^8$ b) $2.72 \cdot 10^5 - 3.1 \cdot 10^3$

<u>Solution:</u>

a) $3.75 \cdot 10^8 + 2.11 \cdot 10^8 = (3.75 + 2.11) \cdot 10^8 = \boxed{5.86 \cdot 10^8}$.

b) $2.72 \cdot 10^5 - 3.1 \cdot 10^3 =$ Notice they do not have the same exponents. So, we must reduce to common exponent. And we can choose:

$$2.72 \cdot 10^5 - 3.1 \cdot 10^3 = \left| 2.72 \cdot 10^5 = 2.72 \cdot 10^2 \cdot 10^3 = 272 \cdot 10^3 \right| = 272 \cdot 10^3 - 3.1 \cdot 10^3 =$$

$$= (272 - 3.1) \cdot 10^3 = 268.9 \cdot 10^3 = \boxed{2.689 \cdot 10^5}.$$

Or

$$2.72 \cdot 10^5 - 3.1 \cdot 10^3 = \left| 3.1 \cdot 10^3 = 3.1 \cdot 10^{-2} \cdot 10^5 = 0.031 \cdot 10^5 \right| = 2.72 \cdot 10^5 - 0.031 \cdot 10^5 =$$

$$= (2.72 - 0.031) \cdot 10^5 = \boxed{2.689 \cdot 10^5}.$$

Exercises

14. Work out:

a) $4.732 \cdot 10^5 + 4.1 \cdot 10^3$ b) $2.575 \cdot 10^4 - 7.1 \cdot 10^3$

15. Write in scientific notation:

a) 1,230,000,000,000,000 b) 0.000000000001230

16. Write in decimal notation:

a) 5.213×10^7 b) 4.723×10^{-6} c) 4.2×10^8 d) 8.7091×10^{-4}

Exercises

17. Planck´s constant, 6.626176×10^{-34}, is one of the lowest positive numbers used in Physics. Written in decimal notation it is 0.000 ... 6626176, how many zeros are there between decimal point and the first significant digit?

18. Work out, expressing the result in scientific notation:

a) $7.14 \cdot 10^4 + 6.234 \cdot 10^4$ b) $8.27310^4 - 1.496 \cdot 10^6$ c) $1.273 \cdot 10^{-3} \cdot 4.197 \cdot 10^5$

d) $(5.12 \cdot 10^3) : (1.28 \cdot 10^{-5})$ e) $\dfrac{5.24 \cdot 10^{-5} - 3.7 \cdot 10^{-7}}{2.645 \cdot 10^3 + 3.9}$ f) $\dfrac{(2.6 \cdot 10^{-1}) \cdot (7.2 \cdot 10^3)}{(3.8 \cdot 10^7) \cdot (6.5 \cdot 10^{-4})}$

2. Roots

Till now, we have calculated and worked with powers. In this part of the unit, we are going to work with the function "root", which is reciprocal to a power.

For example:

POWERS	ROOTS
A square has a side of 5 cm. What is its area? $A = s^2 = 5^2 = 25\ cm^2$.	The area of a square is 49 cm^2. What is the length of its side? s = 7 cm, because $7^2 = 49$. This is a root, a square root: $\sqrt{49} = 7 \Leftrightarrow 7^2 = 49$
A cube has a side of 4 m. What is its volume? $V = s^3 = 4^3 = 64\ m^3$.	The volume of a cube is cube 8 cm^3. What is the length of its side? s = 2 cm, because $2^3 = 8$. This is a root, a cubic root: $\sqrt[3]{8} = 2 \Leftrightarrow 2^3 = 8$

Root of a real number ***a***, written $\sqrt[n]{a}$ or $\mathbf{a}^{1/n}$, where ***n*** is a natural number, is another real number, ***b***, so that $\sqrt[n]{a} = b \text{ or } a^{1/n} \Leftrightarrow b^n = a$

In this expression, ***n*** is named ***index*** and ***a*** is named ***radicand***.

Number of roots

a) Even index:

- If radicand is a **positive** number, there are **two opposite roots**.

For example, squared root of 25 is 5 and (-5).

- If radicand is **0**, its root is **0.**

- If radicand is a **negative** number, it has **no roots,** because any number to an even power, is always a positive number.

For example, (-4) has no roots, because there is not any number b, so that b2 = (-4).

b) Odd index:

In this case, all numbers have **an only root**: **Positive** if radicand is **positive**, and **negative** if radicand is **negative**, and **null** if radicand is **null**.

For example, $\sqrt[3]{8} = 2$, $\sqrt[3]{-8} = -2$ and $\sqrt[3]{0} = 0$

Following chart resumes above information:

Index: n	Radicand: a	Number of roots
Even	a > 0	2 opposite roots
	a = 0	1 null root
	a < 0	No roots
Odd	a > 0	1 positive root
	a = 0	1 null root
	a < 0	1 negative root

Exercises

19. Write as a power:

a) $\sqrt[3]{7}$ b) $\sqrt[4]{a^3}$ c) $\sqrt{10}$ d) $\sqrt[5]{(a+2)^2}$ e) $\sqrt[6]{b^5}$ f) $\sqrt[3]{n^2}$

g) $\sqrt[4]{m^3}$ h) $\sqrt[3]{x}$ i) $\dfrac{1}{\sqrt[3]{x}}$ j) $\sqrt[3]{6xy}$ k) $4\sqrt[3]{x^2}$

20. Calculate, if possible, the value of the following expressions:

a) $4^{\frac{1}{2}}$ b) $-4^{\frac{1}{2}}$ c) $(-4)^{\frac{1}{2}}$ d) $(-8)^{\frac{1}{3}}$ e) $0^{\frac{1}{3}}$

21. Write as a root:

a) $25^{\frac{1}{3}}$ b) $12^{\frac{1}{4}}$ c) $a^{\frac{3}{5}}$ d) $\left(\frac{1}{2}\right)^{\frac{-2}{3}}$ e) $b^{\frac{2}{7}}$ f) $2^{\frac{3}{2}}$

g) $2^{\frac{1}{2}}$ h) $(a^2+b^2)^{\frac{1}{2}}$ i) $(a-b)^{\frac{-1}{2}}$ j) $2x^{\frac{3}{5}}$ k) $(2x^3y)^{\frac{2}{5}}$ l) $(8x)^{\frac{-2}{3}}$

m) $(-8)^{\frac{-1}{5}}$ n) -3

22. Write as a power:

a) $(\sqrt[3]{3})^6$ b) $(\sqrt[3]{4})^8$ c) $(\sqrt{3})^2$ d) $(\sqrt[5]{31})^5$ e) $(\sqrt{n^2})^7$ f) $(\sqrt[3]{m^3})^5$

g) $(\sqrt[5]{m^2})^3$ h) $(\sqrt[4]{5})^2$ i) $\sqrt[3]{\sqrt{128}}$

Equivalent roots. Simplification of roots

In roots having a power as radicand, ***index*** and ***exponent*** can be ***multiplied*** or ***divided*** by a ***same number***, in the same way you are used to do it with fractions.

When we divide index and exponent by a same number, we are ***simplifying*** the root.

As we have seen above, a root can be written as a power, being its exponent a fraction. So, the same operations we use with fractions, can be used with roots (simplifying, ordering, …….).

Example: Simplify: a) $\sqrt[6]{8}$ b) $\sqrt[12]{625}$ c) $\sqrt[18]{a^{12}}$

Solution (as root):

a) $\sqrt[6]{8} = \sqrt[6]{2^3} = \sqrt[2]{2^1} = \sqrt{2}$, where we have divided by 3.

b) $\sqrt[12]{625} = \sqrt[12]{5^4} = \sqrt[3]{5}$, where we have divided by 4.

c) $\sqrt[18]{a^{12}} = \sqrt[3]{a^2}$, where we have divided by HCF(18, 12) = 6.

Solution (as power):

a) $\sqrt[6]{8} = \sqrt[6]{2^3} = 2^{\frac{3}{6}} = 2^{\frac{1}{2}} = \sqrt{2}$, where numerator and denominator have been divided by 3.

b) $\sqrt[12]{625} = \sqrt[12]{5^4} = 5^{\frac{4}{12}} = 5^{\frac{1}{3}} = \sqrt[3]{5}$, where numerator and denominator have been divided by 4.

c) $\sqrt[18]{a^{12}} = a^{\frac{12}{18}} = a^{\frac{2}{3}} = \sqrt[3]{a^2}$, where numerator and denominator have been divided by HCF(18, 12) = 6.

Exercises

23. Simplify, if possible, the following roots:

a) $\sqrt[12]{a^{10}}$ b) $\sqrt[3]{a^{12}}$ c) $\sqrt[15]{3^{10}}$ d) $\sqrt[4]{64}$ e) $\sqrt[4]{3^2}$

f) $\sqrt[9]{27}$ g) $\sqrt[8]{5^4}$

24. Write as powers with the same index:

a) $\sqrt{2}$; $\sqrt[3]{3}$; $\sqrt[4]{4}$ b) $2^{\frac{3}{2}}$; $8^{\frac{1}{2}}$; $\sqrt{8}$; $\sqrt[5]{5}$ c) $\sqrt[4]{8}$; $\sqrt[3]{5}$; $\sqrt[12]{600}$

d) $\sqrt[3]{n^2}$; $\sqrt[4]{m^3}$; $\sqrt[6]{n^5}$ e) $\sqrt{a^3}$; $\sqrt[4]{m}$; $\sqrt[5]{n^5}$; $\sqrt[10]{a^3}$

Comparing roots

> To ***compare*** two or more roots, they must have the ***same index***. If they do not have it, we will have to transform our roots into another ones, equivalent to them and having the same index. (Common index will be LCM of original ones).

Example: List in an increasing order the roots $\sqrt[4]{2^3}$, $\sqrt{2}$ and $\sqrt[6]{2^5}$

Solution: Write them as powers: $\sqrt[4]{2^3} = 2^{\frac{3}{4}}$; $\sqrt{2} = 2^{\frac{1}{2}}$; $\sqrt[6]{2^5} = 2^{\frac{5}{6}}$ As all bases are 2, we only have to arrange exponent in an increasing. order: $\frac{3}{4}, \frac{1}{2}$ *and* $\frac{5}{6}$ Remember, we must transform these fractions in others, equivalent and having as denominators, LCM(4, 2, 6) = 12: $\frac{9}{12}, \frac{6}{12}$ *and* $\frac{10}{12}$.

So, as $\frac{1}{2} < \frac{3}{4} < \frac{5}{6}$, then, $\sqrt{2} < \sqrt[4]{2^3} < \sqrt[6]{2^5}$.

Exercises	
	25. Simplify the following roots: a) $\sqrt[6]{5^3}$ b) $\sqrt[15]{2^{12}}$ c) $\sqrt[10]{a^8}$ d) $\sqrt[12]{a^4 \cdot b^8}$ e) $\sqrt[8]{x^2 y^2}$ f) $\sqrt[4]{x^{12}}$

Laws of roots

Law	Example	Description
$\sqrt[n]{a} \cdot \sqrt[n]{b} = \sqrt[n]{a \cdot b}$	$\sqrt{2} \cdot \sqrt{3} = \sqrt{6}$	**Product** of two roots having the same index is a root having the common index and whose radicand is the product of radicands.
$\sqrt[n]{a} : \sqrt[n]{b} = \sqrt[n]{a : b}$	$\sqrt[5]{12} : \sqrt[5]{4} = \sqrt[5]{3}$	**Quotient** of two roots having the same index is a root having the common index and whose radicand is the quotient of radicands.

$\left(\sqrt[n]{a}\right)^m = \sqrt[n]{a^m}$	$\left(\sqrt[5]{4}\right)^2 = \sqrt[5]{16}$	**Power** of a root is another root having the same index and whose radicand is the power of the original radicand.
$\sqrt[m]{\sqrt[n]{a}} = \sqrt[n\cdot m]{a}$	$\sqrt{\sqrt[3]{8}} = \sqrt[6]{8}$	**Root** of a root is another root having the same radicand and whose index is the product of the original index.

These laws can be easily demonstrated by applying the power rules:

a) $\sqrt[n]{a}\cdot\sqrt[n]{b} = a^{\frac{1}{n}}\cdot b^{\frac{1}{n}} = (a\cdot b)^{\frac{1}{n}} = \sqrt[n]{a\cdot b}$

b) $\sqrt[n]{a} : \sqrt[n]{b} = a^{\frac{1}{n}} : b^{\frac{1}{n}} = (a : b)^{\frac{1}{n}} = \sqrt[n]{a : b}$

c) $\left(\sqrt[n]{a}\right)^m = \left(a^{\frac{1}{n}}\right)^m = a^{\frac{m}{n}} = \left(a^m\right)^{\frac{1}{n}} = \sqrt[n]{a^m}$

d) $\sqrt[m]{\sqrt[n]{a}} = \left(a^{\frac{1}{n}}\right)^{\frac{1}{m}} = a^{\frac{1}{n\cdot m}} = \sqrt[n\cdot m]{a}$

Exercises

26. Calculate:

a) $\sqrt{3}\cdot\sqrt{5}\cdot\sqrt{7}$ b) $\sqrt[3]{4}\cdot\sqrt[3]{5}\cdot\sqrt[3]{2}$ c) $\sqrt[5]{12}\div\sqrt[5]{6}$ d) $\sqrt{4}\div\sqrt{2}$

e) $\sqrt[3]{\sqrt{5}}$ f) $\sqrt[6]{\sqrt[3]{6}}$ g) $\sqrt{5}\cdot\sqrt{6}$ h) $\sqrt{2a}\cdot\sqrt{a}$

i) $\sqrt{18}\div\sqrt{50}$ j) $\sqrt{12}\cdot\sqrt{\frac{3}{4}}\cdot\sqrt{\frac{12}{5}}\cdot\sqrt{\frac{15}{4}}$

Introducing and extracting factors from roots

These laws allow us to introduce or extract factors from roots.

Example: Introduce factors into the following roots:

Solution: a) $2\sqrt{3}$ b) $2\sqrt[3]{5}$

a) $2\sqrt{3} = \sqrt{2^2}\cdot\sqrt{3} = \sqrt{4}\cdot\sqrt{3} = \boxed{\sqrt{12}}$

b) $2\sqrt[3]{5} = \sqrt[3]{2^3}\cdot\sqrt[3]{5} = \sqrt[3]{8}\cdot\sqrt[3]{5} = \boxed{\sqrt[3]{40}}$

radicands

Example: Extract from the following roots: a) $\sqrt{200}$ b) $\sqrt[3]{250}$

Solution:

a) $\sqrt{200}=\sqrt{2^3\cdot5^2}=\sqrt{2\cdot2^2\cdot5^2}=2\cdot5\cdot\sqrt{2}=\boxed{10\sqrt{2}}$

b) $\sqrt[3]{250}=\sqrt[3]{5^3\cdot2}=\sqrt[3]{5^3}\cdot\sqrt[3]{2}=\boxed{5\sqrt[3]{2}}$

Exercises

27. Introduce in the roots:

a) $5\sqrt{3}$ b) $2\sqrt[3]{4}$ c) $3\sqrt[4]{2}$ d) $2\sqrt[3]{2}$ e) $2\sqrt{5}$ f) $7\sqrt{a}$

g) $2a\sqrt{3a}$ h) $x\sqrt{\frac{1}{x}}$ i) $x^3y\sqrt{xy}$ j) $\frac{1}{3}\sqrt[4]{\frac{27}{2}}$ k) $\frac{3}{2}\sqrt{\frac{2}{3}}$ l) $\frac{2}{a}\sqrt{\frac{ax}{2}}$

m) $\frac{3}{2xy}\sqrt{\frac{2xz}{y}}$

28. Extract all possible factors from the following roots:

a) $\sqrt{3^5}$ b) $\sqrt[4]{5^{10}}$ c) $\sqrt[3]{81}$ d) $\sqrt{300}$ e) $\sqrt{18}$ f) $\sqrt{32}$

g) $\sqrt{8}$ h) $\sqrt{75}$ i) $\sqrt{200}$ j) $\sqrt[3]{625}$ k) $\sqrt[4]{32}$ l) $\sqrt[4]{243}$

29. Extract all possible factors from the following roots:

a) $\sqrt{\frac{27}{4}}$ b) $\sqrt[5]{\frac{5x^{10}}{y^8}}$ c) $\sqrt[3]{\frac{8x^4y^3z}{n^6}}$ d) $3\sqrt{8a^3}$ e) $2x^2y\sqrt{x^4y^3}$

f) $\frac{xy^2}{3}\sqrt{27xy^3}$ g) $\sqrt{8}$ h) $\sqrt{12}$ i) $\sqrt[3]{16}$ j) $\sqrt[3]{54}$

k) $\sqrt[5]{64}$ l) $\sqrt{12x^3y^5z^2}$ m) $\sqrt[3]{\frac{8x^4}{81y^6}}$

Like roots

Two or more radicals are said to be ***like radicals*** if they have the *same index* and the *same radicand*.

For example, these radicals are like radicals: $\sqrt[5]{14}$, $8\sqrt[5]{14}$ and $(-2)\sqrt[5]{14}$.

To ***add*** or ***subtract*** radicals, they must be like radicals.

Example: Calculate: a) $\sqrt[5]{14} - 8\sqrt[5]{14} + 10\sqrt[5]{14}$

Solution:

a) $\sqrt[5]{14} - 8\sqrt[5]{14} + 10\sqrt[5]{14} = (1-8+10)\sqrt[5]{14} = \boxed{3\sqrt[5]{14}}$.

b) $6\sqrt[3]{5} - \sqrt[3]{40} = 6\sqrt[3]{5} - \sqrt[3]{2^3 \cdot 5} = 6\sqrt[3]{5} - 2\sqrt[3]{5} = \boxed{4\sqrt[3]{5}}$

c) $\sqrt{2} + \sqrt{8} + \sqrt{18} - \sqrt{32} = \sqrt{2} + \sqrt{2^3} + \sqrt{2 \cdot 3^2} - \sqrt{2^5} = \sqrt{2} + 2\sqrt{2} + 3\sqrt{2} - 4\sqrt{2} = (1+2+3-4)\sqrt{2} = \boxed{2\sqrt{2}}$

Exercises

30. Add the following radicals, extracting previously all possible factors:

a) $\sqrt{45} + \sqrt{20} - \sqrt{500} + \sqrt{80}$

b) $\sqrt{24} - 5\sqrt{6} + \sqrt{486}$

c) $\sqrt[3]{54} - \sqrt[3]{16}$

d) $2\sqrt[3]{81} + \frac{1}{3}\sqrt[3]{3} - \frac{2}{5}\sqrt[3]{24}$

31. Add the following radicals, making them, previously, like radicals:

a) $\sqrt{x} + \sqrt{xy^2} - \sqrt{xy^4} - \sqrt{xy^6}$

b) $\sqrt{5x} + \sqrt{45x} + \sqrt{180x} - \sqrt{80x}$

c) $\sqrt{24xy^2} - 5\sqrt{6x^3} + \sqrt{486x^5y^4}$

Rationalization

Sometimes, when operating with radicals, there are fractions with radicals at the denominator. ***Rationalizing*** consists on finding another equivalent fraction, without radicals at the denominator.

To rationalize a fraction, you must multiply ***numerator*** and ***denominator*** by a same expression that makes radicals at denominator disappear.

- If denominator is $\sqrt{a}$, we will multiply numerator and denominator by $\sqrt{a}$.

- If denominator is $a+\sqrt{b}$, $a-\sqrt{b}$, $\sqrt{a}+\sqrt{b}$ or $\sqrt{a}-\sqrt{b}$ we will multiply numerator and denominator by their **conjugated** (the same expression, changing the central sign).

Example: Rationalize: a) $\frac{5}{\sqrt{3}}$ b) $\frac{3}{2\sqrt{2}}$ c) $\frac{3}{\sqrt{2}-1}$

Solution:

a) $\frac{5}{\sqrt{3}}$ → We must multiply numerator and denominator by $\sqrt{3}$ → $\frac{5\cdot\sqrt{3}}{\sqrt{3}\cdot\sqrt{3}}=\frac{5\cdot\sqrt{3}}{(\sqrt{3})^2}=\boxed{\frac{5\cdot\sqrt{3}}{3}}$

b) $\frac{3}{2\sqrt{2}}$ → We must multiply numerator and denominator by $\sqrt{2}$ → $\frac{3\cdot\sqrt{2}}{2\cdot\sqrt{2}\sqrt{2}}=\frac{3\cdot\sqrt{2}}{2\cdot(\sqrt{2})^2}=\boxed{\frac{3\cdot\sqrt{2}}{4}}$

c) $\frac{3}{\sqrt{2}-1}$ → We must multiply numerator and denominator by $(\sqrt{2}+1)$ →

→ $\frac{3\cdot(\sqrt{2}+1)}{(\sqrt{2}-1)\cdot(\sqrt{2}+1)}=\frac{3\cdot(\sqrt{2}+1)}{(\sqrt{2})^2-1}=\frac{3\cdot(\sqrt{2}+1)}{2-1}=\boxed{3\sqrt{2}+3}$

Exercises

32. Rationalize the following expressions:

a) $\frac{2}{\sqrt{3}}$ b) $\frac{1}{3\sqrt{2}}$ c) $\frac{5}{\sqrt{5}}$ d) $\frac{2\sqrt{6}}{\sqrt{2}}$ e) $\frac{\sqrt{27}}{\sqrt{8}}$

f) $\sqrt{\frac{2}{3}}$ g) $\sqrt{\frac{5}{2}}$ h) $\frac{3}{\sqrt{5}}$

33. Multiply by its conjugated each of the following expressions:

a) $2-\sqrt{x}$ b) $\sqrt{x}-2$ c) $\sqrt{x}+\sqrt{y}$

34. Rationalize the following expressions:

a) $\frac{2}{3-\sqrt{5}}$ b) $\frac{8\sqrt{5}}{\sqrt{7}-\sqrt{3}}$ c) $\frac{9}{\sqrt{x}-\sqrt{y}}$ d) $\frac{1+\sqrt{x}}{1-\sqrt{x}}$

35. Rationalize the following expressions:

a) $\frac{2}{1-\sqrt{3}}$ b) $\frac{5}{\sqrt{7}-\sqrt{2}}$ c) $\frac{\sqrt{2}}{5-\sqrt{2}}$ d) $\frac{\sqrt{3}}{\sqrt{12}+\sqrt{2}}$

Review exercises

Powers

1. Write as an only power:

a) $7^3 \cdot 7^2 =$ b) $8^{10} \cdot 8^{20} \cdot 8^1 =$ c) $4^5 : 4^2 =$ d) $(3^2)^6 =$

e) $10^7 : 10^4 =$ f) $(9^2)^8 =$ g) $(3^{100} : 3^{50}) \cdot 3^2 =$ h) $10^4 \cdot 10^2 \cdot 10^5 =$

i) $(3^3 \cdot 3^6) : 3^7 =$

2. Work out:

a) $(-3)^2 \cdot (-3)^{-3}$ b) $6^{-2} : 6^{-6}$ c) $x^2 : x^{-6}$ d) $x^{-2} : x^6$

e) $(2^{-3})^4$ f) $(5^2)^{-4}$ g) $(x^{-3})^{-2}$ h) $(x^2 y^3)^2$

i) $(x^5 y^{-2}) : (x^6 y^{-1})$ j) $(6x^4 y^2) : (3x^2 y^{-2})$ k) $(4x^{-3} y^3) : (2x^{-1} y^{-1})$ l) $5^{-2} : 5^{-3}$

3. Express as an only power:

$a) 6^{-2} \cdot 6^0 =$

$b) 2^{-4} \cdot 2^6 =$

$c) 3^{-2} \cdot 3^{-2} =$

$d) 5^0 : 5^{-2} =$

$e) 6^{-3} : 6^0 =$

$f) 5^8 : 5^{-4} =$

$g) 6^{-3} : 6^{-7} =$

$h) (4^0)^5 =$

$i) (6^{-4})^0 =$

$j) 5 : 5^{-6} =$

$k) (3^{-1})^4 =$

$l) (2^{-3})^2 =$

$m) 7^{-3} : 7^{-5} =$

$n) 3^{-6} : 3^{-2} =$

$ñ) 2^{-4} : 2^{-1} =$

4. Express as an only power:

a) $5^2 \cdot (5^3 \div 5) =$ d) $(6^{-3} \cdot 36^{-2})^{-1} =$ g) $5 \div 5^{-6} =$

b) $(5^2 \cdot 25^2)^3 =$ e) $((-2)^{12})^3 \cdot (8)^5 =$ h) $7^{-3} : 7^{-5} =$

c) $3^{-6} : 3^{-2} =$ f) $2^{-4} : 2^{-1} =$ i) $3^{-2} : 3^{-6} =$

5. Work out:

a) $6^{-2} \cdot 6^0 =$ d) $5^8 \div 5^{-4} =$ g) $(4^2 \cdot 8^3)^{-2} =$ j) $2^{-4} \cdot 2^6 =$

b) $6^{-3} \cdot 6^0 =$ e) $(3^2)^{-6} =$ h) $(8^{-2} \div 4^{-3})^3 =$ k) $3^{-2} \cdot 3^{-2} =$

c) $6^{-3} \div 6^{-7} =$ f) $(2^{-2})^{-3} =$ i) $(2^3 \cdot 4^{-3})^{-4} =$ l) $5^0 \div 5^{-2} =$

6. Simplify:

a. $\left(\frac{2}{3}\right)^{6}\cdot\left(\frac{2}{3}\right)^{-4}=$ b. $\left(\frac{3}{5}\right)^{7}:\left(\frac{3}{5}\right)^{-2}=$ c. $\left(\frac{1}{3}\right)^{5}\cdot\left(\frac{1}{3}\right)^{-3}=$ d. $\left(\frac{3}{4}\right)^{-2}\cdot\left(\frac{3}{4}\right)^{-5}=$

e. $\left[\left(\frac{4}{5}\right)^{3}\right]^{-2}=$ f. $\left[\left(\frac{2}{3}\right)^{-3}\right]^{2}=$ g. $\left[\left(\frac{2}{3}\right)^{-3}\right]^{-2}=$ h. $\dfrac{\left[\left(\frac{2}{3}\right)^{2}\cdot\left(\frac{2}{3}\right)^{-5}\right]^{-3}}{\left[\left(\frac{2}{3}\right)^{-5}:\left(\frac{2}{3}\right)^{-8}\right]^{-2}}=$

i. $\dfrac{\left(\frac{1}{5}\right)^{-5}:\left(\frac{1}{5}\right)^{-9}}{\left(\frac{1}{5}\right)^{3}\cdot\left(\frac{1}{5}\right)^{-10}:\left(\frac{1}{5}\right)}=$

7. Simplify:

a) $\dfrac{3^{2}\cdot5^{6}\cdot3^{-5}\cdot5^{-2}}{3^{4}\cdot5^{7}}=$ b) $\dfrac{2^{5}\cdot2^{3}\cdot2^{-2}\cdot2^{-7}}{2^{4}\cdot2^{-3}}=$ c) $\dfrac{3^{-5}\cdot3^{5}\cdot3^{-2}}{3^{4}\cdot3^{-6}}=$

d) $\dfrac{5^{7}\cdot5^{-3}\cdot5}{5^{-8}\cdot5}=$ e) $\dfrac{7\cdot7^{-3}\cdot7^{2}}{7^{4}\cdot7^{-1}}$ f) $\dfrac{3^{3}\cdot5^{7}\cdot3^{-1}\cdot5}{5^{11}\cdot3\cdot5^{-2}}=$

g) $\dfrac{2^{4}\cdot2^{-3}\cdot5\cdot7^{3}\cdot5^{2}}{7^{4}\cdot5^{3}\cdot2^{-2}}=$ h) $\dfrac{3^{6}\cdot3^{-4}\cdot5^{3}\cdot7^{3}}{5^{4}\cdot7^{4}\cdot3^{-2}}=$ i) $\dfrac{5\cdot11^{2}\cdot5^{-3}\cdot7}{11^{3}\cdot7^{3}\cdot5^{-2}}$

j) $\dfrac{\left(3^{2}\cdot3^{-2}\cdot5\cdot5^{-4}\right)^{6}}{\left(3\cdot5^{2}\cdot5^{-1}\right)^{2}}=$ k) $\dfrac{\left(5^{3}\cdot3\cdot11^{2}\right)^{2}}{\left(3^{2}\cdot5\cdot11^{-1}\right)^{3}}$

8. Simplify:

a) $\dfrac{6^{2}}{10^{3}}$ b) $\dfrac{8^{3}}{6^{4}}$ c) $\dfrac{9^{2}}{6^{3}}$ d) $\dfrac{20^{2}}{8^{3}}$ e) $\dfrac{70^{2}}{28^{3}}$ f) $\dfrac{12^{4}}{16^{2}}$ g) $\dfrac{14^{3}}{21}$

h) $\dfrac{25^{2}}{15^{2}}$ i) $\dfrac{15^{3}}{30^{2}}$

9. Express as an only fraction:

a) $\left(27^{2}\cdot9^{4}\right)^{2}=$ b) $\left(16^{3}\cdot4^{2}\right)^{5}=$ c) $\left(25\cdot5^{3}\right)^{4}=$ d) $\left(49^{2}\cdot7^{5}\right)^{-3}=$

e) $\left(81^{2}\cdot9^{2}\right)^{3}=$ f) $\left(27^{-2}\cdot3^{-4}\right)^{2}=$ g) $\left(121^{2}\cdot11\right)^{-2}=$

10. Simplify:

a) $\dfrac{49^2 \cdot 9^2}{21^2 \cdot 3} =$ b) $\dfrac{75^2 \cdot 20^{-1}}{25^2 \cdot 4^{-1}} =$ c) $\dfrac{18^2 \cdot 24}{8 \cdot 27} =$ d) $\dfrac{3^2 \cdot 3^0 \cdot 3^{-2}}{3^4 \cdot 3^{-1}} =$

e) $\dfrac{\left(\frac{1}{2}\right)^3 \cdot \frac{1}{2} \cdot \left(\frac{1}{2}\right)^{-2}}{\left(\frac{1}{2}\right)^4 \cdot \left(\frac{1}{2}\right)^{-1}} =$ f) $\dfrac{245 \cdot 7^{-2}}{3^{-2} \cdot 3} =$ g) $\dfrac{3^{-2} \cdot 4^3 \cdot 4^{-2}}{9^{-2} \cdot 2^{-2} \cdot 3^{-3}}$ h) $\dfrac{1^{-8} - 2^{-2} + 8}{2^2 + 2^{-2}}$

i) $\dfrac{6^4 \cdot 9^{-2}}{27^{-3}}$ j) $\dfrac{144^2 \cdot 5^2}{36^4 \cdot 20^3}$ k) $\dfrac{2^2 \cdot 3^4}{9 \cdot 12 \cdot 6} =$ m) $\dfrac{4 \cdot 45 \cdot 24}{3^3 \cdot 2^4 \cdot 5} =$ n) $\dfrac{8 \cdot 27^{-1}}{12^{-1}} =$

Scientific notation

11. Work out:

a) $(7{,}823 \cdot 10^{-5}) \cdot (1{,}84 \cdot 10^{13}) =$ b) $(5{,}24 \cdot 10^6) \cdot (6{,}3 \cdot 10^8) =$ c) $\dfrac{5{,}24 \cdot 10^6}{6{,}3 \cdot 10^{-8}} =$

d) $2{,}35 \cdot 10^8 + 1{,}43 \cdot 10^7 =$ e) $5{,}83 \cdot 10^9 - 7{,}5 \cdot 10^{10} =$ f) $(2 \cdot 10^5) \cdot (1{,}5 \cdot 10^7) =$

g) $(5 \cdot 10^{12}) : (2 \cdot 10^{-3}) =$ h) $(8{,}4 \cdot 10^{11}) : (3{,}2 \cdot 10^{-6}) =$ i) $\dfrac{3 \cdot 10^{-5} + 7 \cdot 10^{-4}}{10^6 - 5 \cdot 10^5} =$

j) $\dfrac{7{,}35 \cdot 10^4}{5 \cdot 10^{-3}} + 3{,}2 \cdot 10^7 =$ k) $(4{,}3 \cdot 10^3 - 7{,}2 \cdot 10^5)^2 =$

12. Transform into scientific notation and calculate:

a) $(75800)^4 : (12000)^2 =$ b) $\dfrac{0{,}000541 \cdot 10318000}{1520000 \cdot 0{,}00302} =$ c) $\dfrac{2700000 - 13000000}{0{,}00003 - 0{,}00015} =$

Roots

13. Work out the following operations:

a) $\sqrt{2} \cdot \sqrt{32}$ b) $\sqrt[3]{3} \cdot \sqrt[3]{9}$ c) $\sqrt{2} \cdot 8^{0.5}$ d) $\sqrt{2} \cdot \sqrt[3]{15}$ e) $\sqrt[3]{2} \cdot \sqrt[5]{3}$ f) $\sqrt[3]{2} \cdot \sqrt[5]{8}$

14. Express as an only power and as an only root:

a) $\left(\sqrt{x}\right)^{1/3}$ b) $\sqrt{x} : \sqrt[3]{x}$ c) $\sqrt{x} \cdot \sqrt[3]{x} \cdot \sqrt[5]{x^2}$ d) $\sqrt[3]{\sqrt[5]{x^2}}$ e) $\dfrac{1}{x\sqrt{x}}$ f) $\dfrac{x}{\sqrt[3]{x}}$

15. Introduce or extract factor from the following roots:

a) $2\sqrt{5}$ b) $4\sqrt[3]{7}$ c) $5\sqrt{3}$ d) $5\sqrt[3]{2}$ e) $\sqrt{45}$ f) $\sqrt[3]{54}$

g) $\sqrt{1200}$ h) $\sqrt[3]{500}$

16. Calculate the following divisions of roots:

a) $\sqrt{15}:\sqrt{3}$ b) $\sqrt{2}:\sqrt[3]{32}$ c) $\sqrt{3}:\sqrt{4}$ d) $\sqrt{8}:\sqrt[4]{2}$ e) $\sqrt[3]{81}:\sqrt[3]{9}$

f) $\sqrt[3]{9}:\sqrt[6]{3}$

17. Without using your calculator, calculate the value of the following roots:

a) $\sqrt{0.25}$ b) $\sqrt[3]{\frac{1}{8}}$ c) $\sqrt[3]{0.216}$ d) $\sqrt{2^{-12}}$ e) $\sqrt[3]{7^6}$

18. Calculate:

a) $\sqrt{3}+3\sqrt{3}+7\sqrt{3}$ b) $6\sqrt{3}-4\sqrt{3}+5\sqrt{3}$ c) $3\sqrt{2}+5\sqrt{2}-8\sqrt{2}$

19. Add the following radicals, making them, previously, like radicals:

a) $\sqrt{7}+\sqrt{28}-\sqrt{63}$ b) $\sqrt{121}+\sqrt{169}-\sqrt{225}$ c) $\sqrt{2}+\sqrt{8}+\sqrt{18}-\sqrt{32}$

d) $\sqrt{5}+\sqrt{45}+\sqrt{180}-\sqrt{80}$ e) $\sqrt{24}-5\sqrt{6}+\sqrt{486}$ f) $\sqrt[3]{54}-2\sqrt[3]{16}$

g) $\sqrt{2}+3\sqrt{18}-\sqrt{32}$ h) $\sqrt{5}+\sqrt{180}-\sqrt{80}$ i) $\sqrt{2}+\sqrt{32}+5\sqrt{8}$

j) $3\sqrt{20}+2\sqrt{5}$ k) $2\sqrt{27}-4\sqrt{12}$ l) $3\sqrt{28}-2\sqrt{27}$

m) $\sqrt[3]{16}-\sqrt[3]{54}$

20. Add the following radicals:

a) $2a\sqrt{2}-\sqrt{8}+3\sqrt{2}$ b) $2a\sqrt{3}-\sqrt{27a^2}+a\sqrt{12}$ c) $(3+a)\sqrt{5}-\sqrt{125}+\sqrt{5a^2}$

d) $4\sqrt{12}-\frac{3}{2}\sqrt{48}+\frac{2}{3}\sqrt{27}+\frac{3}{5}\sqrt{75}$ e) $7\sqrt{54}-3\sqrt{18}+\sqrt{24}-\frac{3}{5}\sqrt{50}-\sqrt{6}$

f) $\sqrt{3}-\sqrt{108}+\sqrt{648}-\sqrt{1875}$ g) $\frac{3\sqrt{3}}{2}-\frac{7\sqrt{108}}{4}+\sqrt{648}-\frac{2}{3}\sqrt{1875}$

21. Rationalize the following expressions:

a) $\frac{2}{\sqrt{3}}$ b) $\frac{3y}{x\sqrt{y}}$ c) $\frac{x\sqrt{y}}{y\sqrt{x}}$ d) $\frac{2\sqrt{12}}{5\sqrt{3}}$ e) $\frac{3\sqrt{2}}{\sqrt{6}}$

22. Rationalize the following expressions:

a) $\frac{1}{\sqrt{2}+\sqrt{3}}$ b) $\frac{1}{\sqrt{2}-\sqrt{3}}$ c) $\frac{2}{\sqrt{2}-1}$ d) $\frac{\sqrt{2}}{3-\sqrt{2}}$ e) $\frac{1-\sqrt{2}}{1+\sqrt{2}}$

23. Rationalize the following expressions:

a) $\frac{\sqrt{5}-2}{3-2\sqrt{5}}$ b) $\frac{3\sqrt{5}-2\sqrt{3}}{2\sqrt{3}+3\sqrt{5}}$ c) $\frac{1}{\sqrt{x}-\sqrt{y}}$ d) $\frac{\sqrt{x}}{2-\sqrt{x}}$ e) $\frac{\sqrt{2y}}{\sqrt{2}-\sqrt{y}}$

24. Calculate the value of the following expressions:

a) $\frac{3}{\sqrt{2}}+\frac{1}{2}$ b) $\frac{3}{\sqrt{7}}-\frac{\sqrt{2}}{4}$ c) $\frac{2}{\sqrt{3}}+\frac{3}{\sqrt{2}}$ d) $\frac{2}{\sqrt{3}}+\frac{1}{2\sqrt{3}}$

Unit 3.- Progressions

1. Progressions of real numbers. General term

In the following drawings, observe the process carrying to the creation of new black triangles. Following the process, how many black triangles will Figure 5 have? And, in general, Figure n?

Figure	Number of triangles
1	$1 = 3^0$
2	$3 = 3^1$
3	$9 = 3^2$
4	$27 = 3^3$

In figure n, the number of black triangles is 3^{n-1}. So, for example, using this expression, we can calculate the number of black triangles that figure 10 will have: $3^{(10-1)} = 3^9 = 19683$.

The ordinated number sequence we have obtained, 1, 3, 9, 27, 81, 243, … is named a ***progression of real numbers*** and the expression 3^{n-1} is known as the ***general term*** of the progression.

- A ***real numbers progression*** $a_1, a_2, a_3, a_4, \ldots, a_n, \ldots$ is an arranged sequence of infinite real numbers.
- Each number in the sequence is named ***term***.
- The numbers 1, 2, 3, ... are named ***index***, and indicate the position of a term in the sequence.
- The term a_n is named ***n-th term*** or ***general rule*** or ***formula***, and it is the expression that allows to know the value of a term of the sequence by only knowing its position in the sequence.

Exercises

1. Write the five first terms of the following sequences:

a) Each term is obtained by adding 7 to the previous term. First one is (-10).

b) The first term is 5. Each term is obtained by multiplying the previous term by 2.

c) First term is 2; second one is 4, and following are the semi-addition of the two previous.

2. Write the five first terms of:

a) $a_n = 2n-1$ b) $a_n = 2n^2 + 1$ c) $a_n = 3 \cdot 2^n$ d) $a_n = n - n^2$

3. Write the general term of the following sequences:

a) 1, 5, 9, 13, 17, ... b) 1, 4, 9, 16, 25, 36, ... c) 2, 4, 8, 16, 32, 64, ...

d) 1, (-3), 9, (-27), 81, (-243), ... e) 170, 120, 70, 20, (-30), (-80), ...

4. Write the five first terms of: a) $a_n = n^3$ b) $b_n = n^2 - 3n + 7$ c) $c_n = \frac{n-3}{n+4}$

5. Write the five first terms of the sequences:

a) $a_n = 3 + 5(n - 1)$ b) $b_n = (n - 1) \cdot (n - 2)$ c) $c_n = 3\left(\frac{1}{2}\right)^{n-1}$ d) $d_n = n^2 - n$

6. Check that the general term of the sequence –1, 1, –1, 1, –1, 1, ... is $s_n = (-1)^n$.

Exercises

7. Find out the general term of the following sequences:

a) $a_n \rightarrow 1, -1, 1, -1, 1, -1, \dots$ b) $b_n \rightarrow 1, -2, 3, -4, 5, -6, \dots$

8. Write the general term of the following sequences:

a) 1, 3, 5, 7, 9,......... b) 3, 4, 5, 6, 7, ... c) 1, 4, 9, 16, 25,.....

d) 2, 5, 10. 17, 26,...... e) 2, 4, 8, 16, 32,........ f) –7, -3, 1, 5, 9........

There are sequences that do not have a general term, and each term is calculated from the previous ones. They are named ***recurring sequences***.

Example: The sequence 1, 1, 2, 3, 5, 8, ... is defined by the recurrence rule $a_n = a_{n-1} + a_{n-2}$.

Exercises

9. Form a recurring sequence with the rule $j_1 = 2;\quad j_2 = 3;\quad j_n = j_{n-2} + j_{n-1}$

10. Build a recurring sequence with the rule $a_1 = 5;\quad a_n = a_{n-1} + n$.

11. Add one new term to each sequence and find out the recurring rule for the following sequences:

a) 1, –4, 5, –9, 14, –23, ... (Difference)

b) 1, 2, 3, 6, 11, 20, ... (Relation of each term with the three previous)

c) 1; 2; 1,5; 1,75; ... (Semi-addition)

d) 1, 2, 2, 1, 1/2, 1/2, 1, ... (Quotient)

2. Arithmetic sequences

Look at the following sequence: 3, 8, 13, 18, 23, 28, 33, 38, **...**

As you can see, each term is obtained by adding 5 to the previous one.

An ***arithmetic sequence*** is made by adding a constant quantity to the previous term. This constant quantity is named ***difference***.

So, an arithmetic sequence is perfectly defined by its ***first term*** (a_1) and its ***difference (d)***.

Example: An arithmetic sequence is, for example, 5, 8, 11, 14, 17, 20, …

Notice that $a_1 = 5$ and $d = 3$.

Now, we are going to find out the general term of this arithmetic sequence.

$a_1 = 5$

$a_2 = 5 + 3 = a_1 + d$

$a_3 = 5 + 6 = 5 + 3 \cdot 2 = a_1 + d \cdot 2$

$a_4 = 5 + 9 = 5 + 3 \cdot 3 = a_1 + d \cdot 3$

$\boxed{a_n = a_1 + d \cdot (n - 1)}$

The general term of an ***arithmetic sequence*** is $\boxed{a_n = a_1 + d \cdot (n - 1)}$

Exercises

12. Find the general term and the a_{28} for the following arithmetic sequences:

a) 2, 5, 8, 11, …;

b) 10, 5, 0, -5, -10, -15, …..

c) 120, 140, 160, 180, …

d) 9, 7, 5, 3, 1, -1, -3, …..

13. Calculate a_{31} for the previous sequences.

14. Write the five first terms and the general term of the following arithmetic sequences:

a) Natural numbers.

b) Natural odd numbers.

c) Natural even numbers.

15. Write the general term of the following sequences:

a) 3, 7, 11, 15, ... b) 6, 4, 2, 0, -2, ...

16. Write the five first terms and the general term of the following arithmetic sequences:

a) $a_1 = 1.5; d = 2$ b) $a_1 = 32; d = -5$

c) $a_1 = 5; d = 0.5$ d) $a_1 = -3; d = -4$

17. Write the general term and calculate a_{50} for:

a) 25, 18, 11, 4, ... b) –13, –11, –9, –7, ...

c) 1,4; 1,9; 2,4; 2,9; ... d) –3, –8, –13, –18, ...

18. Find out the first term and the difference of the following arithmetic sequences:

a) $a_2 = 18; a_7 = -17$ b) $a_4 = 15; a_{12} = 39$

c) Even numbers d) Multiples of 3.

19. What is the position of a term whose value is 56 in the arithmetic sequence given by $a_1 = 8$ and $d = 3$?

20. John is going to run a competition. He starts his training running for 3 km, and every day he is increasing his distance in 1.5 km. How many days must he continue his training to run 21 km?

21. If two terms of an arithmetic sequence are $s_1 = 6$ and $s_3 = 9$, find out the value of the difference and write the general term.

Sum of the n first terms of an arithmetic sequence

Imagine one day you start a training jogging for a popular race. This day, you may run for 5 minutes, and the following days your resistance will increase and increase……

Imagine each day you can run for 5 minutes more than previous. The minutes you run each day are given by an arithmetic sequence, 5, 10, 15, 20, 25,……., whose general term is $a_n = 5 + 5(n - 1)$.

But imagine now you want to know the total quantity of minutes you have run during your 7 days training. This is the concept of *sum of the n first terms*.

In the following lines we are going to see a very easy method to calculate it.

You only have to write twice the sequence of terms you want to add. First one, in its natural sense, and second one, in an inverse sense.

$$\begin{array}{ccccccccccccccc} S_7 & = & 5 & + & 10 & + & 15 & + & 20 & + & 25 & + & 30 & + & 35 \\ S_7 & = & 35 & + & 30 & + & 25 & + & 20 & + & 15 & + & 10 & + & 5 \end{array}$$

Notice that if you add each pair of terms, below and above, you have the same result, 40

$$\begin{array}{ccccccccccccccc} S_7 & = & 5 & + & 10 & + & 15 & + & 20 & + & 25 & + & 30 & + & 35 \\ S_7 & = & 35 & + & 30 & + & 25 & + & 20 & + & 15 & + & 10 & + & 5 \\ \hline 2S_7 & = & 40 & + & 40 & + & 40 & + & 40 & + & 40 & + & 40 & + & 40 \end{array}$$

$$2\cdot S_7 = 7\cdot(5+35) \quad \rightarrow \quad S_7 = \frac{7\cdot(5+35)}{2} \quad \rightarrow \quad S_7 = \frac{7\cdot(a_1+a_7)}{2}$$

So, in general: $\boxed{S_n = \dfrac{n\cdot(a_1+a_n)}{2}}$

The sum of the n first terms of an arithmetic sequence is $S_n = \dfrac{n\cdot(a_1+a_n)}{2}$

Example: Calculate the addition of the 500 first even numbers.

Solution:

The sequence of even numbers is 2, 4, 6, 8, 10, ... and it is an arithmetic sequence where $a_1 = 2$ and $d = 2$.

The term $a_{500} = a_1 + (500 - 1)d = 2 + 499 \times 2 = 1.000$

Therefore, the sum of the 500 first values is

$$S_{500} = \frac{500 \cdot (a_1 + a_{500})}{2} = \frac{500 \cdot (2 + 1000)}{2} = \frac{500 \cdot 1002}{2} = \boxed{250,500}.$$

Exercises

22. Calculate the sum of the first 100 terms of the arithmetic sequence whose $a_1 = 7$ y $d = 5$.

23. Calculate the sum of the first 16 terms of the arithmetic sequence given by

$a_n = 5 + 4 \cdot (n-1)$

24. The dose of a drug was 100 mg the first day, and 5 less each day, during 12 days. How many mg of the drug will the person have taken at the end of his treatment?

25. Calculate the sum of all the odd numbers less than 100.

35. a) If $a_1 = 5$ and $a_2 = 7$, calculate a_{40} and S_{40}.

b) If $b_1 = 5$ and $b_2 = 12$, calculate S_{32}.

26. In a sequence, $c_1 = 17$ and $c_5 = 9$, find out the sum S_{20}.

27. Sum $1 + 2 + 3 + 4 + \ldots + n$ of the n first naturals *is* 60 378. Calculate the value of n.

3. Geometric sequences

Look at the following sequences:

2, 6, 18, 54, 162, ...;   2, 1, 1/2, 1/4, 1/8, ...;   3, -3, 3, -3, 3, -3, ...

Notice that, in all of them, each term has been obtained by multiplying the previous one by a constant number. These sequences are named *geometric sequences*. The constant number is named ratio (r).

A ***geometric sequence*** is made by multiplying a constant quantity to the previous term. This constant quantity is named ***ratio***.

So, a geometric sequence is perfectly defined by its ***first term*** (a_1) and its ***ratio (r)***.

Example: A geometric sequence is, for example, 5, 10, 20, 40, 80, 160, ...
Notice that $a_1 = 5$ and $r = 2$.

Now, we are going to find out the general term of this arithmetic sequence.

$a_1 = 5$

$a_2 = 5 \cdot 2 = a_1 \cdot r$

$a_3 = 5 \cdot 2 \cdot 2 = 5 \cdot 2^2 = a_1 \cdot r^2$

$a_4 = 5 \cdot 2 \cdot 2 \cdot 2 = 5 \cdot 2^3 = a_1 \cdot r^3$

$\boxed{a_n = a_1 \cdot r^{n-1}}$

The general term of a ***geometric sequence*** is $\boxed{a_n = a_1 \cdot r^{n-1}}$

Exercises

28. The two first terms of a geometric sequence are 5 and 10. Write down the 3 following terms and the ratio.

29. In a geometric sequence, s_1=100 and s_3=400. Calculate the 5 first terms.

30. In a bacteria population, the number of bacteria is duplicated every day. If the first day there were 4 bacteria, find out:

a) the general term of the geometric sequence.

b) the number of bacteria existing in one month.

31. Find out the general term of the sequence: 3, 6, 12, 24, 48, ...

32. Write the 5 first terms of the following geometric sequences:

a) $a_1 = 0.3$; $r = 2$ b) $a_1 = -3$; $r = ½$ c) $a_1 = 200$; $r = -0.1$ d) $a_1 = 1/81$; $r = 3$

33. Find the general term of the following sequences:

a) 20; 8; 3.2; 1.28; … b) 40, 20, 10, 5, …

c) 6; –9; 13.5; –20.25; … d) 0.48; 4.8; 48; 480; …

34. Find the general term of the following sequences:

a) $a_1 = 1/81$; $a_3 = 1/9$ b) $a_2 = 0.6$; $a_4 = 2.4$

c) $a_3 = 3$; $r = 1/10$ d) $a_4 = 20.25$; $r = -1.5$

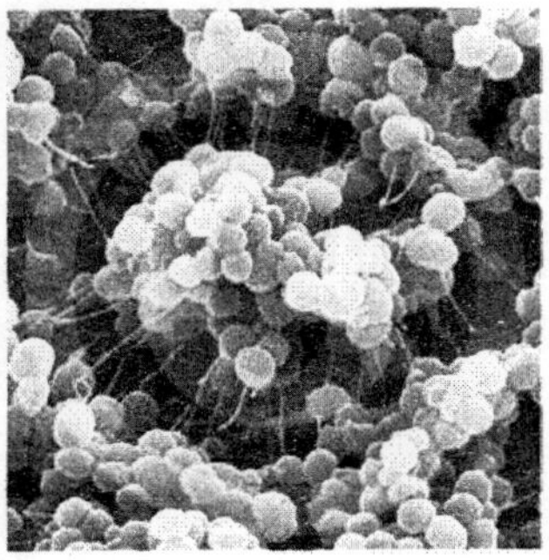

35. Certain kind of bacteria reproduces by bipartition each quarter of an hour. How many bacteria will be produced by each, after 6 hours?

36. Build a geometric sequence whose first term is 125 and ratio = 0.4.

37. In a geometric sequence, a_1 = 0.625 and a_3 = 0.9. Calculate r and the six first terms.

Sum of the n first terms of a geometric sequence

We are going to name $a_1, a_2, a_3, \ldots, a_n, \ldots$ a geometric sequence with ratio r and we are going to calculate the sum of its n first terms.

$$S_n = a_1 + a_2 + a_3 + \ldots\ldots\ldots + a_{n-2} + a_{n-1} + a_n$$

Now, we are going to multiply S_n by r:

$$r \cdot S_n = r \cdot a_1 + r \cdot a_2 + r \cdot a_3 + \ldots\ldots\ldots + r \cdot a_{n-2} + r \cdot a_{n-1} + r \cdot a_n =$$

$$= a_2 + a_3 + a_4 + \ldots\ldots\ldots + a_{n-1} + a_n + r \cdot a_n$$

And, now, we are going to calculate the difference $r \cdot S_n - S_n$:

$$\begin{array}{rcccccccccccccc} r\,S_n & = & & & a_2 & + & a_3 & + & a_4 & + & \cdots & + & a_{n-1} & + & a_n & + & a_n r \\ S_n & = & & a_1 & + & a_2 & + & a_3 & + & a_4 & + & \cdots & + & a_{n-1} & + & a_n & \\ \hline r\,S_n - S_n & = & - & a_1 & & & & & & & & & & & & + & a_n r \end{array}$$

$$r \cdot S_n - S_n = r \cdot a_n - a_1 = r \cdot a_1 \cdot r^{n-1} - a_1 = a_1 \cdot r^n - a_1$$

$$(r-1) \cdot S_n = a_1 \cdot r^n - a_1$$

$$S_n = \frac{a_1 \cdot r^n - a_1}{r-1} = \frac{a_1 \cdot (r^n - 1)}{r-1} = \boxed{a_1 \cdot \frac{r^n - 1}{r-1}}$$

The sum of the n first terms of a geometric sequence is $a_1 \cdot \dfrac{r^n - 1}{r-1}$

Example: Knowing that in a geometric sequence, the terms $a_1 = 3$ and $a_5 = 1875$, calculate the sum of the 5 first terms.

Solution:

First of all, we will calculate the ratio. As $a_5 = a_1 \cdot r^4$, $1875 = 3 \cdot r^4$, $r^4 = 625$; $r = 5$.

The 5 first terms are 3, 15, 75, 375, 1.875, and their sum is 2,343.

Using the expression, $S_n = \dfrac{a_1 \cdot r^n - a_1}{r-1} \rightarrow S_5 = \dfrac{3 \cdot 5^5 - 3}{5-1} = \boxed{2{,}343}$.

Exercises

38. Calculate the sum of the ten first terms of the following geometric sequences:

a) $a_1 = 5; r = 1.2$ b) $a_1 = 5; r = -2$

39. Calculate the sum of the five first terms of a geometric sequence in which $a_1 = 1\ 000$ and $a_4 = 8$.

40. Using the given expression, calculate 3 + 6 + 12 + 24 + 48 + 96 + 192 + 384.
Sol.: 765.

41. Calculate the sum of the ten first terms of a geometric sequence in which $a_1 = 8.192$ y $r = 2.5$.

42. Using the given expression, calculate the sum $1 + 3 + 9 + 27 + \dots + 3^7$.

Advanced material: Sum of the infinite terms of a geometric sequence with /r/<1

Geometric sequences with /r/ < 1, the absolute value of their terms is decreasing. And, moreover, when n becomes relatively large, their values are almost zero.

Example: Calculate the four first terms of the geometric sequence $a_n = 10\cdot(0.5)^n$ and notice that the absolute value of its terms is decreasing.

Solution: $a_n = 5;\ 2.5;\ 1.25;\ 0.625,\ \dots$

As its values are decreasing, till having values very closed to zero, these last terms are insignificant when calculating the sum. So, in these cases, it is useful to calculate the sum of the infinite values, and it is used the expression:

$$\boxed{S_n = \frac{a_1}{1-r}}$$

Example: In a geometric sequence with $a_1 = 25$ and $r = 0.5$, calculate the sum of its infinite terms.

Solution: $S_n = \frac{a_1}{1-r} = \frac{25}{1-0.5} = \frac{25}{0.5} = \boxed{50}$.

Exercises

43. In a geometric sequence with $a_1 = 8$ and $r = 0.75$, calculate the sum of its infinite terms.

44. In a geometric sequence with $a_1 = 30$ and $r = -0.2$, calculate the sum of its infinite terms.

45. In a geometric sequence with $a_4 = 10$ and $a_6 = 0.4$, calculate:

a) The ratio, r.

b) The first term, a_1.

c) The 8th term, a_8.

d) The sum of its 8 first terms, S_8.

e) The sum of its infinite terms.

Review exercises

Sequences

1. Complete the following sequences till the 10th term:

a) 5 , 9 , 13, 17 , 21 , 25 , 29 , 33 , 37 , 41

b) 22 , 19 , 16 , 13 , 10 , ___ , ___ , ___ ,

c) 3 , 6 , 12 , 24 , 48 , 96 , 192 , 384 , 728 , 1456

d) 1 , 4 , 9 , 16 , 25 , ___ , ___ , ___ , ___

e) 3 , 4 , 7 , 11 , 18 , ___ , ___ , ___ , ___ , ___

2. Write the five first terms of the sequence $a_n = n^2 - 2n$

3. Write the five first terms of the sequence $b_n = 2^{n-1}$

4. Write the five first terms of the sequence $c_n = \dfrac{1}{2n}$

Arithmetic sequences

5. Complete the following sequences to make them arithmetic sequences, and indicate the difference in each case:

a) 5 , 9 , 13 , 17 , 21 , 25 , 29 , ... d = 4

b) 8 , ___ , 11 , ___ , 14 , ___ , ___ , ... d =

c) -10 , -6 , -2 , 2 , 6 , 10 , 14 , ... d = -4

d) 32 , 22 , 12 , 2 , -8 , -18 , -28 , ... d = -10

e) 1 , 6 , 11 , 16 , 21 , 26 , 31 , ... d = 5

6. Write the ten first terms of an arithmetic sequence whose 6th term is 16 and whose difference is 2.5. Calculate its general term.

7. Knowing the 5th term of an arithmetic sequence, 7, and its difference, 4, write its five first terms and its general term. $a_5 = 7 = a_1 + (5-1)4$

$d = 4$

8. Write the ten first terms of an arithmetic sequence whose 4th term is 11 and whose difference is (-3). What is the general term of this sequence?

9. First term of an arithmetic sequence is 12 and 5th term is 18. Write its eight first terms and its general term.

10. Fifth term of an arithmetic sequence is 9 and 9th term is 11. Write its ten first terms and its general term.

11. First term of an arithmetic sequence is (-8) and 6th term is 13. Write its ten first terms and its general term.

Sum of the first terms of an arithmetic sequence

12. In an arithmetic sequence, 12th term is 42 and the difference is 2. Calculate the sum of its 20 first terms.

13. What is the sum of the 50 first terms of an arithmetic sequence if the 7th term is 23 and 8th one is 27?

14. My son George wants to save money to buy a bike that costs 44.90 euros. I am going to give him 15 cents the first day he has a good behavior, 30 cents the second day, 45 the third, etc. George thinks if he has a good behavior for 4 weeks, he will be able to buy the bike. Is this true?

15. How many terms of an arithmetic sequence have been added if the sum is 165, $a_1 = 5$, and the las added term is 25?

Geometric sequences

16. Complete the following sequences to make them geometric sequences, and indicate the ratio in each case:

a) 5 , 15 , ___ , ___ , ___ , 1215 , ___ , ... r =

b) 3, ___ , 12 , ___ , 48 , ___ , ___ , ... r =

c) ___ , ___ , ___ , ___ , 64 , 256 , 1024 , ... r =

d) 2000 , 1000 , 500 , ____ , ____ , ____ , ____ , ... r =

17. What are the six first terms of a geometric sequence if 4th term is 13.5 and its ratio is 3? Write the general term.

18. Ratio of a geometric sequence is 0.4 and the third term is 100,000. Write the 8 first terms and its general rule.

19. A biologist has studied the reproduction of bacteria T–52B4 and the result is that a population of these bacteria increases a 20% each day when environmental conditions are appropriate. If the third day there were 14,062,500 bacteria, write the evolution of its population during the 7 first days.

20. Second term of a geometric sequence is 512 and fifth one is 8. Write the 6 first terms and the general term.

$a2 = 512$

$a5 = 8$

21. 4th term of a geometric sequence is 13.5 and 6th one is 30.375. Write the 6 first terms and the general term.

Sum of the first terms of a geometric sequence

22. Calculate the sum of the 12 first terms of a geometric sequence if its first term is 8 and the third is 72.

23. Calculate the sum of the 20 first terms of a geometric sequence knowing its first term, 6.25 and the forth, 10.8.

24. I work as a waiter. In January, my boss has promised me to double the tips each month if I work properly. If in January he has paid me 6 € from tips,

a) How much will I receive in August? b) How much will I receive during the whole year?

c) Are these promises a good idea?

Advanced material: Sum of all the terms of a geometric sequence when /r/ < 1.

25. How much is the sum of all the terms of a geometric sequence whose first term is 1000 and whose 4th term is 125?

26. Calculate this infinite sum: 200 + 20 + 2 + 0´2 + 0´02 + 0´002 + 0´0002 + ...

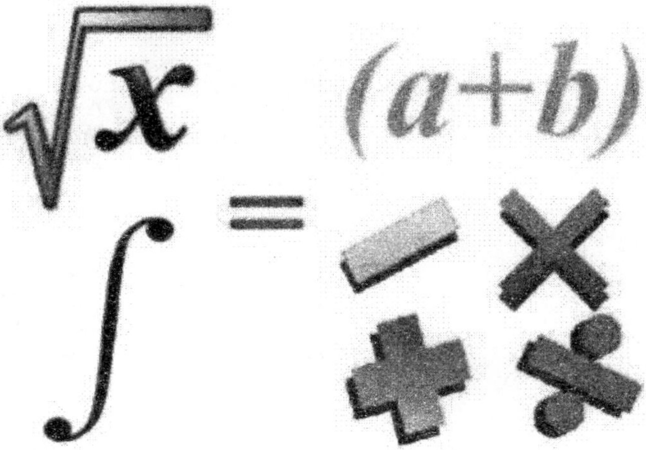

Unit 4.- Algebraic expressions

1. Algebraic expressions

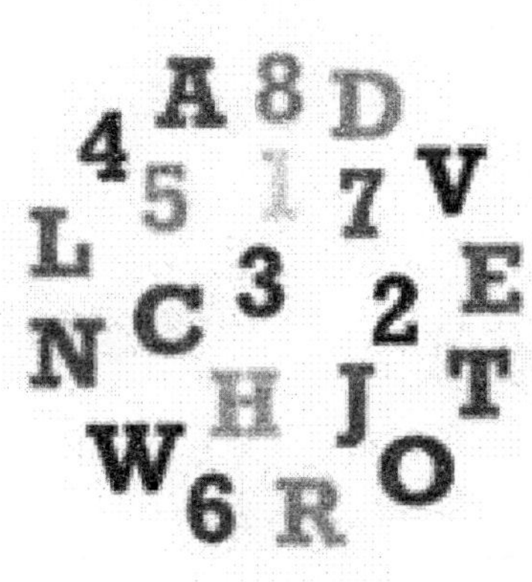

An algebraic expression is a combination of numbers, variables (algebraic quantities) and arithmetic operators: +, –, x, ÷.

For example: 2ax + 5b – 3x2y

Algebraic expressions

$2x$	2 multiplied by x
$5x^2$	5 multiplied by x squared
$3xy$	3 multiplied by x multiplied by y

Terms

In the following algebraic expressions, the terms are separated by "+" or "–" signs.

Examples:

$4x^2 - 2x + 1$ has three terms

$6x + 7$ has two terms

2. Monomials and operations

A monomial consists of the product of a known number **(coefficient)** by one or several letters with exponents that must be constant and positive whole numbers **(literal part)**.

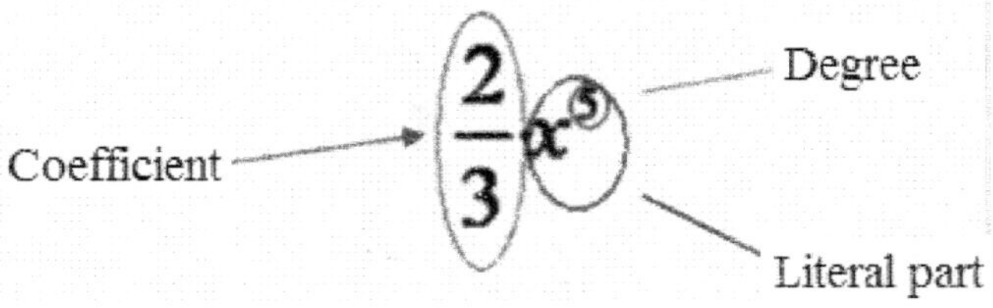

Two or more monomials are said to be **like monomials** if they have the same literal part.

Examples: The expressions $21x^2$, $8x^3$, $\frac{2}{5}x^2$, $-3x^5$ are monomials, and $21x^2$ and $\frac{2}{5}x^2$ are *like monomials*.

Exercises

1. Write a monomial with unknown x that verifies the following conditions:

a) Degree 2 and coefficient (-4).

b) Degree 0 and coefficient 1.

c) A monomial being like to this $4x^3$ and with coefficient 3.

d) Degree 3 and coefficient 1/3.

2.1. Addition and subtraction of monomials

Two monomials can only be added or subtracted if they are **like monomials**. So, they must have the same literal part. In this case, we add or subtract the coefficients and we **leave the literal part unchanged**. When the literal part is different, the addition is left indicated.

The result of the **addition** or **subtraction** two **like monomials** is another like monomial whose coefficient is the sum or subtraction of the coefficients.

$$ax^n \pm bx^n = (a \pm b)x^n.$$

Exercises

2. Calculate:

a) $3x^2 + 4x^2 - 5x^2$

b) $6x^3 - 2x^3 + 3x^3$

c) $x^5 + 4x^5 - 7x^5$

d) $-2x^4 + 6x^4 + 3x^4 - 5x^4$

e) $7x + 9x - 8x + x$

f) $2y^2 + 5y^2 - 3y^2$

g) $3x^2y - 6x^2y + 5x^2y$

h) $4xy^2 - xy^2 - 7xy^2$

i) $2a^6 - 3a^6 - 2a^6 + a^6$

j) $ab^3 + 3ab^3 - 5ab^3 + 6ab^3 - 4ab^3$

k) $7xy^2z - 2xy^2z + xy^2z - 6xy^2z$

l) $-x^3 + 5x - 2x + 3x^3 + x + 2x^3$

m) $x^4 + x^2 - 3x^2 + 2x^4 - 5x^4 + 8x^2$

n) $3a^2b - 5ab^2 + a^2b + ab^2$

o) $\frac{7}{3}x^2 + \frac{4}{3}x^2$

p) $12x^5 - x^5 - 4x^5 - 2x^5 - 3x^5$

q) $\frac{7}{4}x^5 - \frac{1}{4}x^5$

r) $x^2y^2 - 5x^2y^2 - (3x^2y^2 - 4x^2y^2) - 8x^2y^2$

s) $x^2 + \frac{x^2}{3}$

t) $x^2 + x^2$

u) $\frac{1}{2}x^3 - \frac{5}{2}x^3 + \frac{3}{2}x^3$

v) $-(ab^3 + a^3b) - 3a^3b + 5ab^3 - (a^3b - 2ab^3)$

w) $7x^3 - \frac{1}{2}x^3 - \frac{5}{2}x^3 + 2x^3 + \frac{3}{2}x^3$

x) $-x + x^2 + x^3 + 3x^2 - 2x^3 + 2x + 3x^2$

y) $2a^2b + 5a^2b - \frac{2}{3}a^2b - a^2b + \frac{a^2b}{2}$

2.2. Product and quotient of two monomials

To **multiply** two or more monomials, we must multiply, separately, their coefficients and literal parts.

Remember the method to multiply powers having the same base:

$$x^m \cdot x^n = x^{(m+n)}.$$

Example: Multiply: a) $5x^2 \cdot 7x^3$ b) $\frac{3}{2}x^2 \cdot 4x^3$

Solution:

a) $5x^2 \cdot 7x^3 = \left\{ \begin{array}{l} 5 \cdot 7 = 35 \\ x^2 \cdot x^3 = x^{(2+3)} = x^5 \end{array} \right\} = \boxed{35x^5}$

b) $\frac{3}{2}x^2 \cdot 4x^3 = \left\{ \begin{array}{l} \frac{3}{2} \cdot 4 = \frac{3}{2} \cdot \frac{4}{1} = \frac{12}{2} = 6 \\ x^2 \cdot x^3 = x^{(2+3)} = x^5 \end{array} \right\} = \boxed{6x^5}$

To **divide** two or more monomials, we must divide, separately, their coefficients and literal parts.

Remember the method to divide powers having the same base:

$$x^m : x^n = x^{(m-n)}.$$

Example: Divide: a) $15x^6 : 3x^2$ b) $\frac{3}{2}x^7 : \frac{9}{8}x^3$

Solution:

a) $15x^6 : 3x^2 = \left\{ \begin{array}{l} 15 : 3 = 5 \\ x^6 : x^2 = x^{(6-2)} = x^4 \end{array} \right\} = \boxed{5x^4}$

b) $\frac{3}{2}x^7 : \frac{9}{8}x^3 = \left\{ \begin{array}{l} \frac{3}{2} : \frac{9}{8} = \frac{3}{2} \cdot \frac{8}{9} = \frac{3 \cdot 2 \cdot 2 \cdot 2}{2 \cdot 3 \cdot 3} = \frac{2 \cdot 2}{2} = \frac{4}{3} \\ x^7 : x^3 = x^{(7-3)} = x^4 \end{array} \right\} = \boxed{\frac{4}{3}x^4}$

Exercises

3. Calculate:

a) $3x^2 \cdot 4x^3$

b) $2x^3 \cdot 4x^3 \cdot 3x^3$

c) $x^3 \cdot x^3$

d) $-2x^4 \cdot 3x^3$

e) $7x \cdot (-8x^2)$

f) $(-3y^2) \cdot (-2y^3)$

g) $3x^2y \cdot 6xy^3$

h) $\frac{3}{4}x^2 \cdot \frac{5}{2}x^3$

i) $4a^3b^2 \cdot a^2b \cdot 7ab$

j) $-\frac{1}{2}a^3 \cdot \frac{5}{3}a^4$

k) $2a^6 \cdot 3a^6 \cdot 2a^6$

l) $\frac{2}{5}x^3 \cdot \left(-\frac{3}{2}x\right)$

m) $ab^3 \cdot (-3a^2b) \cdot 5a^3b$

n) $x^2 \cdot \frac{1}{3}x^5$

o) $-ab^2c^3 \cdot (-3a^2bc) \cdot 3abc$

p) $(6x^4) : (2x^2)$

q) $\frac{12a^6}{3a^3}$

r) $15x^4 : (-3x)$

s) $\frac{-14x^7}{7x^2}$

t) $-8x^4 : (-4x^3)$

u) $\frac{5x^7y^3}{x^2y}$

v) $(-18x^4) : (6x^3)$

w) $\frac{-12a^5b^4c^6}{2a^3b^2c}$

x) $2x^4 \cdot 6x^3 : 4x^2$

y) $\frac{3a^5b \cdot (-12a^4b^2)}{4a^3b^2}$

z) $27x^4 : (-9x^3) \cdot (-2x^2)$

4. What is the degree of the following monomials?

a) $-5xy^2z^3$ b) $11xy^2$ c) -12

5. Work out the sum of the following monomials:

a) $5x + 3x^2 - 11x + 8x - x^2 + 7x$

b) $6x^2y - 13x^2y + 3x^2y - x^2y$

c) $2x - 5x^2 + 3x + 11y + 2x^3$

d) $3yz^3 + y^3z - 2z^3y + 5zy^3$

Exercises

6. Work out the following products of monomials:

a) $(3x) \cdot (5x^2)$ b) $(-3x^2) \cdot (4x^3)$ c) $\left(\frac{2}{3}x^3\right) \cdot (-6x)$

d) $\left(\frac{2}{9}x^2\right) \cdot \left(-\frac{3}{5}x^3\right)$ e) $(7xy^2) \cdot (2y)$ f) $(5xyz) \cdot (-3x^2z)$

7. Work out:

a) $8x^4 + 5x^4$ b) $7x^3 - 9x^3$ c) $-2x^2 + 4x^5 + 3x^2 - x^5$ d) $4x^7 - 2x + 5x - 4$

e) $3x^4 \cdot (-2x^3)$ f) $(-3x^8) \bullet (-6x^3)$

8. Work out the following combined operations with monomials:

a) $15x^5 - 3x^3 \cdot 4x^2$ (Sol: $3x^5$)

b) $2x^3 + 4x^3 \cdot 5x - 2x \cdot (-x^2)$ (Sol: $4x^3+20x^4$)

c) $3a \cdot ab - 2a^2 \cdot (-4b) - 8 \cdot (2a^2b)$ (Sol: $-5a^2b$)

d) $3x^2 + 4x^2 - 2x^2 \cdot (-3x) - [(4x^3 + x^2 - 2x \cdot (x^2)]$ (Sol: $4x^3+6x^2$)

e) $-3xy^2 - (-4x \cdot 7y^2) + (8x^2y^3 : 2xy)$ (Sol: $29xy^2$)

f) $(-y^2) \cdot (-2y^2) - 5y \cdot (-2y^3) + 3y^3 \cdot (-4y)$ (Sol: 0)

g) $(3x^3 \cdot 6x - 2x^2 \cdot x^2) : (4x^2 \cdot 3x^2 - 8x \cdot x^3)$ (Sol: 4)

h) $3x^5 - \frac{4}{3}x^2 \cdot \frac{3}{2}x^3$ (Sol.: x^5)

i) $4a^2b \cdot (-ab^2) \cdot 5ab - 8a^4b^4$ (Sol: $-28a^4b^4$)

j) $a^5 + \frac{5}{6}a^3 \cdot \frac{3}{5}6^2$ (Sol.: $\frac{3a^5}{2}$)

k) $5x^6 - 2x^6 \cdot 3x^6 : (-2x^6)$ (Sol: $8x^6$)

l) $\left(-\frac{7}{3}x^3\right) \cdot \left(-\frac{4}{7}x\right) + \frac{2}{3}x^4$ (Sol.: $2x^4$)

m) $2ab \cdot (-a^3b) + [ab^2 \cdot (-3a^2b)] - 5a^3b \cdot ab + ab \cdot a^2b^2$ (Sol: $-7a^4b^2-2a^3b^3$)

3. Polynomials and their operations

Remember we cannot add or subtract unlike monomials. We must leave these operations indicated. These sums or subtractions of monomials are named polynomials. For example, we have the polynomial $9x^3 + 8x^2 + 6x + 12$.

A ***polynomial*** in the unknown x is an algebraic expression given by the sum or subtraction of two or more monomials in the same unknown.

- We name ***term*** to each of the monomials in the polynomial. The monomial with degree 0 is named ***independent term***. So, ***binomial*** is a polynomial with two monomials, e.g. $3x + 5$; ***trinomial*** is a polynomial with three monomials, e.g. $3x^2 + 5x + 2$.

· We name ***degree*** of a polynomial to the highest of the degrees of the monomials in it.

· A polynomial is ***complete*** when it has all the terms, from the term with degree 0 to the term with the highest degree.

· A polynomial is ***ordinated*** when the degrees of the terms are increasing or decreasing. Usually, monomials are ordinated in a decreasing order.

Polynomials are named as $A(x)$, $B(x)$, etc. indicating in the parentheses the unknown. So, the following expressions are polynomials.

Polynomial	Number of terms	Independent term	Degree	Complete	Ordinated
$A(x) = 5x^4 - 3x^2 + x - 1$	4	-1	4	Not	Yes
$B(x) = 2 - 5x^2 + 4x$	3 (trinomial)	2	2	Yes	Not
C(x) 3x - 7	2 (binomial)	-7	1	Yes	Yes

3.1. Evaluation of a polynomial

Numerical value of a polynomial $P(x)$ when $x = a$, is the obtained result when substituting x by a and operating. It is named $P(a)$.

Example 3: Evaluate the polynomial $P(x) = 2x^3 + 3x^2 + 4$ when $x = 5$

Solution: $P(5) = 2 \times 5^3 + 3 \times 5^2 + 4 = 2 \text{ x } 125 + 3 \text{ x } 25 + 4 = 250 + 75 + 4 = \boxed{329}$.

Notice that the numerical value of a polynomial when $x = 0$ is its independent term.

Example: Evaluate the polynomial $P(x) = 2x^3 + 3x^2 + 4$ when $x = 0$

Solution: $P(0) = 2 \times 0^3 + 3 \times 0^2 + 4 = 0 + 0 + 4 = \boxed{4}$.

Exercises

9. Evaluate the following polynomials:

a) $P(x) = x^2 + x + 1$,	when $x = 2$	(Sol: 7)
b) $P(x) = x^2 + x + 1$,	when $x = -2$	(Sol: 3)
c) $P(x) = 2x^2 - x + 2$,	when $x = 3$	(Sol: 17)
d) $P(x) = 2x^2 - x + 2$,	when $x = -2$	(Sol: 12)
e) $P(x) = -x^2 - 3x + 4$,	when $x = 4$	(Sol: –24)
f) $P(x) = -x^2 + 3x + 4$,	when $x = -1$	(Sol: 0)
g) $P(x) = x^3 + 3x^2 + 1$,	when $x = 0$	(Sol: 1)
h) $P(x) = x^3 - 4x^2 + x + 3$,	when $x = -3$	(Sol: –63)
i) $P(x) = x^4 - 4x^2 - 1$,	when $x = 2$	(Sol: –1)
j) $P(x) = -x^3 - 3x^2 - x + 2$,	when $x = -4$	(Sol: 22)

Exercises

10. Calculate:

a) Given $P(x) = x^2 + 2x + k$, calculate k so that $P(2) = 6$ (Sol: $K = -2$)

b) Given $P(x) = x^2 - kx + 2$, calculate k so that $P(-2) = 8$ (Sol: $K=1$)

c) Given $P(x) = kx^3 - x^2 + 5$, calculate k so that $P(-1) = 1$ (Sol: $K=3$)

3.2. Sum and subtraction of polynomials

Adding or subtracting polynomials consists simply in adding or subtracting their like monomials. Given two polynomials $A(x) = x^3 - 2x^2 - 7x - 4$ and $B(x) = -x^4 + 8x^2 + 7x + 2$, we can calculate A(x) + B(x) by writing one below the other, so that we collocate in the same column the like monomials.

After that, we add the like monomials.

$$\begin{array}{rl} A(x) = & \qquad x^3 - 2x^2 - 7x - 4 \\ B(x) = & -x^4 \qquad + 8x^2 + 7x + 2 \\ \hline A(x) + B(x) = & -x^4 + x^3 + 6x^2 \qquad - 2 \end{array}$$

Opposite polynomial of a given polynomial is another one that has the same monomials but with opposite coefficients (opposite signs).

If $B(x) = -x^4 + 8x^2 + 7x + 2$, its opposite polynomial is $-B(x) = x^4 - 8x^2 - 7x - 2$.

Do you agree that $3 - 2 = 3 + (-2)$?

So, in the same way, we can calculate $A(x) - B(x) = A(x) + [-B(x)]$:

$$\begin{array}{rl} A(x) = & \qquad x^3 - 2x^2 - 7x - 4 \\ -B(x) = & x^4 \qquad - 8x^2 - 7x - 2 \\ \hline A(x) - B(x) = & x^4 + x^3 - 10x^2 - 14x - 6 \end{array}$$

Another method to add or subtract polynomials is the application of the sign´s rule for addition or subtraction of like monomials.

First of all, we remove the parentheses:

- If it is preceded by a + sign, we conserve the original signs.
- If it is preceded by a – sign, we change the sign inside the parenthesis.

We are using this method to calculate again A(x) + B(x) and A(x) - B(x):

$$A(x) + B(x) = (x^3 - 2x^2 - 7x - 4) + (- x^4 + 8x^2 + 7x + 2) = x^3 - 2x^2 - 7x - 4 - x^4 + 8x^2 + 7x + 2 =$$

$$= \boxed{-x^4+x^3+6x^2-2}$$

$$A(x) - B(x) = (x^3 - 2x^2 - 7x - 4) - (- x^4 + 8x^2 + 7x + 2) = x^3 - 2x^2 - 7x - 4 + x^4 - 8x^2 - 7x-2 =$$

$$= \boxed{x^4+x^3-10x^2-14x-6}$$

Exercises

11. Given the following polynomials: $P(x) = 2x^3 - 3x^2 + 4x - 2$

$Q(x) = x^4 - x^3 + 3x^2 + 4$

$R(x) = 3x^2 - 5x + 5$

$S(x) = 3x - 2$

Calculate:

a) P(x) + Q(x) (Sol: x^4+x^3+4x+2)

b) P(x) + R(x) (Sol: $2x^3-x+3$)

c) P(x) + S(x) (Sol: $2x^3-3x^2+7x-4$)

d) S(x) + P(x) (Sol: ídem)

e) P(x) + P(x) (Sol: $4x^3-6x^2+8x-4$)

f) Q(x) – S(x) (Sol: $x^4-x^3+3x^2-3x+6$)

g) Q(x) + R(x) (Sol: $x^4-x^3+6x^2-5x+9$)

Exercises	**11.** (Continue...)	
	h) P(x) – R(x)	(Sol: $2x^3–6x^2+9x–7$)
	i) Q(x) + S(x)	(Sol: $x^4–x^3+3x^2+3x+2$)
	j) P(x) – S(x)	(Sol: $2x^3–3x^2+x$)
	k) S(x) – P(x)	(Sol: $–2x^3+3x^2–x$)
	l) P(x) – P(x)	(Sol: 0)
	m) R(x) – S(x)	(Sol: $3x^2–8x+7$)
	n) P(x) – Q(x) + R(x)	(Sol: $–x^4+3x^3–3x^2–x–1$)
	o) Q(x) – [R(x) + S(x)]	(Sol: $x^4–x^3+2x+1$)
	p) S(x) – [R(x) – Q(x)]	(Sol: $x^4 – x^3 + 8x –3$)

3.3. Product of two polynomials

To **multiply** two polynomials, we must multiply each monomial of the first one by all the monomials of the second one, or vice versa.

For example, given the polynomials $A(x) = x^3 + 2x^2 - x + 2$ y $B(x) = 4x - 2$, we calculate $A(x) \times B(x)$:

$$
\begin{array}{rr}
A(x) = & x^3 + 2x^2 - x + 2 \\
\times\ B(x) = & 4x - 2 \\
\hline
 & -2x^3 - 4x^2 + 2x - 4 \\
 & 4x^4 + 8x^3 - 4x^2 + 8x \quad \\
\hline
A(x) \cdot B(x) = & 4x^4 + 6x^3 - 8x^2 + 10x - 4
\end{array}
$$

As we have already said, we can multiply two polynomials by applying the distributive property. We mean:

$$A(x) \times B(x) = (x^3 + 2x^2 - x + 2) \times (4x - 2) = x^3(4x - 2) + 2x^2(4x - 2) - x(4x - 2) + 2(4x - 2) =$$

$$= 4x^4 - 2x^3 + 8x^3 - 4x^2 - 4x^2 + 2x + 8x - 4 = \boxed{4x^4 + 6x^3 - 8x^2 + 10x - 4}$$

Exercises

12. Calculate:

a) $(3x^4 - 2x^3 + 2x^2 + 5)\cdot 2x^2$ (Sol.: $6x^6 - 4x^5 + 4x^4 + 10x^2$)

b) $(-2x^5 + 3x^3 - 2x^2 - 7x + 1)\cdot(-3x^3)$ (Sol.: $6x^8 - 9x^6 + 6x^5 + 21x^4 - 3x^3$)

c) $\left(\frac{2}{3}x^3 - \frac{3}{2}x^2 + \frac{4}{5}x - \frac{5}{4}\right)\cdot 12x^2$ (Sol.: $8x^5 - 18x^4 + \frac{48}{5}x^3 - 15x^2$)

d) $\left(\frac{1}{2}ab^3 - a^2 + \frac{4}{3}a^2b + 2ab\right)\cdot 6a^2b$ (Sol.: $3a^3b^4 - 6a^4b + 8a^4b^2 + 12a^3b^2$)

13. Given the following polynomials:

$P(x) = 2x^3 - 3x^2 + 4x - 2$

$Q(x) = x^4 - x^3 + 3x^2 + 4$

$R(x) = 3x^2 - 5x + 5$

$S(x) = 3x - 2$

Calculate the following products:

a) $P(x) \cdot R(x)$ (Sol: $6x^5-19x^4+37x^3-41x^2+30x-10$)

b) $P(x) \cdot S(x)$ (Sol: $6x^4-13x^3+18x^2-14x+4$)

c) $S(x) \cdot P(x)$ (Sol: Ídem)

d) $P(x) \cdot P(x)$ (Sol: $4x^6-12x^5+25x^4-32x^3+4x^2-8x+4$)

e) $Q(x) \cdot S(x)$ (Sol: $3x^5-5x^4+11x^3-6x^2+12x-8$)

f) $[Q(x)]^2$ (Sol: $x^8-2x^7+7x^6-6x^5+17x^4-8x^3+24x^2+16$)

g) $R(x) \cdot S(x)$ (Sol: $9x^3-21x^2+25x-10$)

h) $[R(x)]^2$ (Sol: $9x^4-30x^3+55x^2-50x+25$)

Exercises

14. Work out:

a) $(x^3 + 2) \cdot [(4x^2 + 2) - (2x^2 + x + 1)]$ (Sol: $2x^5 - x^4 + x^3 + 4x^2 - 2x + 2$)

b) $(x^3 + 2) \cdot (4x^2 + 2) - (2x^2 + x + 1)$ (Sol: $4x^5 + 2x^3 + 6x^2 - x + 3$)

c) $(2x^2 + x - 2)(x^2 - 3x + 2) - (5x^3 - 3x^2 + 4)$ (Sol: $2x^4 - 5x^3 - x^2 + 8x - 4$)

d) $(x^2 - 3x + 2) \cdot [(5x^3 - 3x^2 + 4) - (2x^2 + x - 2)]$ (Sol: $5x^5 - 20x^4 + 12x^3 - x^2 - 20x + 12$)

e) $2x^2 + x - 2 - (x^2 - 3x + 2) \cdot (5x^3 - 3x^2 + 4)$ (Sol: $-5x^5 + 18x^4 - 19x^3 + 4x^2 + 13x - 10$)

15. Given the following polynomials:

$P(x) = 2x^3 - 3x^2 + 4x - 2$

$Q(x) = x^4 - x^3 + 3x^2 + 4$

$R(x) = 3x^2 - 5x + 5$

$S(x) = 3x - 2$

Calculate

a) $[P(x) + Q(x)] \cdot R(x)$ (Sol: $3x^6 - 2x^5 + 17x^3 - 14x^2 + 10x + 10$)

b) $[Q(x) - R(x)] \cdot S(x)$ (Sol: $3x^5 - 5x^4 + 2x^3 + 15x^2 - 13x + 2$)

c) $[P(x) + Q(x) - S(x)] \cdot R(x)$ (Sol: $3x^6 - 2x^5 + 8x^3 + 7x^2 - 15x + 20$)

d) $[P(x) - Q(x)] \cdot [R(x) + S(x)]$ (Sol: $-3x^6 + 11x^5 - 28x^4 + 33x^3 - 44x^2 + 24x - 18$)

e) $P(x) + 2Q(x)$ (Sol: $2x^4 + 3x^2 + 4x + 6$)

f) $P(x) - 3\,[Q(x) + R(x)]$ (Sol: $-3x^4 + 5x^3 - 21x^2 + 19x - 29$)

g) $P(x) - 2Q(x) + 3R(x)$ (Sol: $-2x^4 + 4x3 - 11x + 5$)

h) $2\,P(x) \cdot Q(x) - R(x)$ (Sol: $4x^7 - 10x^6 + 26x^5 - 26x^4 + 44x^3 - 39x^2 + 37x - 21$)

i) $Q(x) \cdot [2R(x) - 3S(x)]$ (Sol: $6x^6 - 25x^5 + 53x^4 - 73x^3 + 72x^2 - 76x + 64$)

j) $-[Q(x) + 2R(x)] \cdot S(x)$ (Sol: $-3x^5 - x^4 - 29x^3 + 48x^2 - 62x + 28$)

3.4. Quotient of a polynomial by a monomial

To divide a polynomial by a monomial, we divide each term of the polynomial by the monomial. So, each term needs to be divisible by the monomial. In other case, division is not possible.

Quotient of a polynomial by a monomial, (if possible) is another polynomial whose terms are obtained by dividing each term of the polynomial by the monomial.

Example: Divide $\dfrac{12x^3-9x^2+3x}{3x}=\dfrac{12x^3}{3x}-\dfrac{9x^2}{3x}+\dfrac{3x}{3x}=\boxed{4x^2-3x+1}$

Exercises

16. Calculate the following quotients:

a) $\dfrac{6x^3+4x^2-8x}{2x}$

b) $\dfrac{4x^5-2x^4+4x^3-8x^2+16x}{-2x}$

c) $\dfrac{15x^5-10x^4+5x^2}{5x^2}$

d) $\dfrac{10x^7-12x^5-24x^3}{6x^3}$

17. Calculate the following quotients:

a) $(8x^8 - 6x^4 - 4x^3) : (-4x^3)$

b) $\dfrac{-3x^4+6x^3-12x^2}{3x}$

c) $(-18x^3yz^3) : (6xyz^3)$

d) $\dfrac{-12x^9+2x^5-x^4}{4x^4}$

3.5. Quotient of two polynomials

Quotient of two polynomials follows the same rule than quotient of two whole numbers.

In whole numbers, given two numbers, named *dividend* and *divisor*, we find another two numbers, named *quotient* and *remainder*, such that: $D = d \times q + r$ con $r < d$:

The same is applied to polynomials:

Calculating the quotient of two polynomials, named ***dividend*, D(x)**, and ***divisor*, *d*(x)**, consists on finding another two polynomials, named **quotient, q(x)**, and **remainder, r(x)**, such that:

$$D(x) = d(x) \times q(x) + r(x); \text{ degree } r(x) < \text{degree } d(x)$$

You are going to be shown, with an example, the method to divide two polynomials,

D(x) = $3x^3 - 2x^2 + 5x - 12$ and d(x) = $x^2 - 3x - 5$.

$$3x^3 - 2x^2 + 5x - 12 \quad \underline{|x^2 - 3x - 5}$$

Dividing two polynomials

Step 1: Write both polynomials in a decreasing order of their terms. If dividend is not a **complete** polynomial, leave gaps in the places of the non-existing terms.	$3x^3 - 2x^2 + 5x - 12 \quad \underline{\lvert x^2 - 3x - 5}$
Step 2: Calculate the quotient of the first term of dividend divided by the first term of the divisor. $3x^3 : x^2 = 3x$	$3x^3 - 2x^2 + 5x - 12 \quad \underline{\lvert x^2 - 3x - 5}$ $3x$
Step 3: obtained term of the quotient is multiplied by the whole divisor, and the resulting polynomial is subtracted to the dividend. In this way, we are obtaining the first *partial remainder.*	$3x^3 - 2x^2 + 5x - 12 \quad \underline{\lvert x^2 - 3x - 5}$ $\underline{-3x^3 + 9x^2 + 15x} \quad 3x$ $7x^2 + 20x$

Step 4: We write down the following term of the dividend after the remainder. We divide the first term of the first partial dividend by the first term of the divisor. We continue this process till obtaining a remainder with degree lower than the divisor´s degree.	$\begin{array}{rl} 3x^3-2x^2+5x-12 & \underline{\lvert x^2-3x-5} \\ \underline{-3x^3+9x^2+15x} & 3x+7 \\ 7x^2+20x-12 & \\ \underline{-7x^2+21x+35} & \\ 41x+23 & \end{array}$

Notice that **degree q(x) = degree D(x) – degree d(x)** and it is a general rule.

$$\begin{array}{l|l} D(x) & d(x) \\ \hline r(x) & q(x) \end{array} \rightarrow \text{And verifies:} \quad \begin{array}{l} D(x) = d(x) \cdot Q(x) + r(x) \\ \text{degree } r(x) < \text{degree } d(x) \\ \text{degree } q(x) = \text{degree } D(x) - \text{degree } d(x) \end{array}$$

Exercises

18. Calculate the quotient and remainder of the division $(6x^3 - 5x^2 - 14x - 2) : (2x + 1)$. Check that:

a) $D(x) = d(x) \times q(x) + r(x)$

b) degree $r(x)$ < degree $d(x)$

c) degree $q(x)$ = degree $D(x)$ - degree $d(x)$

19. Work out the following divisions, indicating the quotient and the remainder:

a) $(8x^4 - 4x^2 + 7x) : 2x^2$

b) $(-3x^4 + 3x^2 + 2x - 2) : (x^2 + 2x + 4)$

c) $(3x^4 - x^3 - 6x^2 + 29x - 9) : (3x - 1)$

d) $(7x^4 - 2x^3 + 2x^2 - 5x + 3) : (2x^2 - x + 1)$

20. Work out the following divisions:

a) $(x^4-x^3+7x^2+x+15) : (x^2+2)$

b) $(2x^5-x^3+2x^2-3x-3) : (2x^2-3)$

c) $(6x^4-10x^3+x^2+11x-6) : (2x^2-4x+3)$

d) $(x^3+2x^2+x-1) : (x^2-1)$

e) $(8x^5-16x^4+20x^3-11x^2+3x+2) : (2x^2-3x+2)$

f) $(x^4+3x^3-2x+5) : (x^3+2)$

g) $(x^5-2x^4+3x^2-6) : (x^4+1)$

h) $x^2 : (x^2+1)$

Exercises

21. Work out the following divisions:

a) $(3x^6+2x^4-3x^2+5) : (x^3-2x+4)$ (Sol: $Q(x)=3x^3+8x+12$; $R(x)=13x^2-56x+53$)

b) $x^8 : (x^2+1)$ (Sol: $Q(x)=x^6-x^4+x^2-1$; $R(x)= 1$)

c) $(x^3-4x^2+5x-8) : (x-2)$ (Sol: $Q(x)=x^2-2x+1$; $R= -6$)

d) $(2x^5+3x^2-6) : (x+3)$ (Sol: $Q(x)=2x^4-6x^3+18x^2-51x+153$; $R(x)= -465$)

e) $(x^4-7x^3+8x^2-2) : (x-1)$ (Sol: $Q(x)=x^3-6x^2+2x+2$; exact division)

f) $(3x^5-x^4+8x^2-5x-2) : (x^2-x+1)$ (Sol: $Q(x)=3x^3+2x^2-x+5$; $R(x)= x-7$)

g) $(5x^4-2x^3+x-7) : (x^2-1)$ (Sol: $Q(x)=5x^2-2x+20$; $R(x)= -7x+73$)

h) $(4x^5-3x^3+5x^2-7) : (2x^2-3x+5)$ (Sol: $Q(x)=2x^3+3x^2-2x-8$; $R(x)= -14x+33$)

i) $(9x^3+3x^2- 7x+2) : (3x^2+5)$ (Sol: $Q(x)=3x+1$; $R(x)= -22x-3$)

j) $(4x^4-3x^2+5x-7) : (2x^2+x-3)$ (Sol: $Q(x)=2x^2-x+2$; $R(x)= -1$)

k) $(4x^5+3x^3-2x^2+5) : (2x^2-x+3)$ (Sol: $Q(x)=2x^3+x^2-x-3$; $R(x)= 14$)

l) $(6x^4+5x^2-3x+8) : (3x^3-2x-3)$ (Sol: $Q(x)=2x$; $R(x)= 9x^2+3x+8$)

m) $(4x^4+2x^3-3x^2+5x-1) : (2x^2-3)$ (Sol: $Q(x)=2x^2+x+3/2$; $R(x)= 8x+7/2$)

3.6. Quotient of a polynomial by (x – a)

One of the most frequent situation when dividing polynomials is having as divisor a binomial like *(x – a)*, being *a* a real number. For example, $(5x^4 - 4x^2 + 2x - 9) : (x - 4)$, where a = 4.

These divisions can be calculated in the general way, as above, but there is a more comfortable and quick method, named *Ruffinis´s Rule*. Riffini´s rule uses only the coefficients of the dividend.

Ruffini´s Rule: We are going to show it with the example $x^4 - 8x^2 + 3x - 1 : (x - 2)$.

Ruffini's Rule

Step 1: Write the coefficients of dividend. You must list them in a decreasing order of degree. Write zero in the places of non-existing terms. Write the independent term of the divisor, <u>changing its sign</u>.	$\begin{array}{r\|rrrrr} & 1 & 0 & -8 & 3 & -1 \\ 2 & & & & & \\ \hline & & & & & \end{array}$
Step 2: Write down the first coefficient of the dividend.	$\begin{array}{r\|rrrrr} & 1 & 0 & -8 & 3 & -1 \\ 2 & & & & & \\ \hline & 1 & & & & \end{array}$
Step 3: Multiply the opposite of the independent term of the divisor, by this number you have just written down, and write it into the box, under the second term of the dividend.	$\begin{array}{r\|rrrrr} & 1 & 0 & -8 & 3 & -1 \\ 2 & & 2 & & & \\ \hline \times & 1 & & & & \end{array}$
Step 4: Add in vertical.	$\begin{array}{r\|rrrrr} & 1 & 0 & -8 & 3 & -1 \\ & & + & & & \\ 2 & & 2 & & & \\ \hline & 1 & 2 & & & \end{array}$
Step 5: Repeat the product of 2 by the last number you wrote down in the quotient zone.	$\begin{array}{r\|rrrrr} & 1 & 0 & -8 & 3 & -1 \\ 2 & & 2 & 4 & & \\ \hline \times & 1 & 2 & & & \end{array}$

Final steps: Continue this process till the end. Last obtained number in the quotient line is "remainder". It is usually written in a box.

$$\begin{array}{r|rrrrr} & 1 & 0 & -8 & 3 & -1 \\ 2 & & 2 & 4 & -8 & -10 \\ \hline & 1 & 2 & -4 & -5 & \boxed{-11} \end{array} \leftarrow \text{Remainder}$$

So, solution is: Quotient: $x^3 + 2x^2 - 4x - 5$; Remainder: (-11)

Important note: Remember that, in general, degree q(x) = degree D(x) – degree d(x). In this case, as divisor (x – a) has degree = 1, degree q(x) = degree D(x) – 1.

Example: Determine the value of *k* that makes $4x^3 + 16x^2 + k$ divisible by $x + 3$.

Solution:

The remainder of the division $(4x^3 + 16x^2 + k) : (x + 3)$ must be zero. Applying the Ruffini's rule, we have:

	4	16	0	k
–3		–12	–12	36
	4	4	–12	$k + 36 = 0$

From the equation $k + 36 = 0$, we obtain that $k = (-36)$.

Root of a polynomial: We say a is a ***root*** of a polynomial P(x) if quotient $\frac{P(x)}{x-a}$ is an exact division (remainder = 0).

Example: Check if 2 is a root of the polynomial P(x) = $3x^2$ - 12x + 12

Solution: We must divide ($3x^2$ - 12x + 12) : (x – 2), and we do it by using the Ruffini's rule:

	3	- 12	12
2		6	- 12
	3	- 6	0

As remainder = 0, 2 is a root of the given polynomial.

Exercises

22. Work out the following divisions using the Ruffini's rule:

a) $(x^3-4x^2+5x-8) : (x-2)$ (Sol: $Q(x)=x^2-2x+1$; $R=-6$)

b) $(x^4-7x^3+8x^2-2) : (x-1)$ (Sol: $Q(x)=x^3-6x^2+2x+2$; Exact)

c) $(2x^4+3x^3-4x^2+x-18) : (x-2)$ (Sol: $Q(x)=2x^3+7x^2+10x+21$; $R=24$)

d) $(x^3+x^2+x+1) : (x+1)$ (Sol: $Q(x)=x^2+1$; Exact)

e) $(2x^4+x^3-2x^2-1) : (x+2)$ (Sol: $Q(x)=2x^3-3x_2+4x-8$; $R=15$)

f) $(2x^5+3x^2-6) : (x+3)$ (Sol: $Q(x)=2x^4-6x^3+18x^2-51x+153$; $R=-465$)

g) $(3x^4-10x^3-x^2-20x+5) : (x-4)$ (Sol: $Q(x)=3x^3+2x^2+7x+8$; $R=37$)

h) $(2x^4-10x+8) : (x+2)$ (Sol: $Q(x)=2x^3-4x^2+8x-26$; $R=60$)

i) $(10x^3-15) : (x+5)$ (Sol: $Q(x)=10x^2-50x+250$; $R=-1265$)

j) $(x^3+2x^2+3x+1) : (x-1)$ (Sol: $Q(x)=x^2+3x+6$; $R=7$)

k) $(x^4-2x^3+x^2+3x+1) : (x-2)$ (Sol: $Q(x)=x^3+x+5$; $R=11$)

l) $(2x^4-7x^3+4x^2-5x+6) : (x-3)$ (Sol: $Q(x)=2x^3+5x^2+x-2$; Exact)

m) $(x^5+1) : (x-1)$ (Sol: $Q(x)=x^4+x^3+x^2+x+1$; $R=2$)

23. Answer the following questions:

a) Calculate the value of m, so that when dividing $(x^3 - 3x^2 - mx + 12)$ by $(x - 3)$, remainder is 9.

b) Find n so that the division $(2x^4 + x^3 + 5x^2 + 5x - n) : (x - 1)$ is exact.

c) Find the value of a that makes the polynomial $x^3 - ax^2 - ax + 1$, divisible by $x - 1$.

d) Find m y n so that the polynomial $P(x) = x^3 - 2x^2 + (m - 1)x + n$ verifies $P(0) = 2$ and it is divisible by $x - 1$.

4. Polynomial or Remarkable identities. Special products

When working on polynomials, there are three identities that are going to be very useful for you to save time. They are named *polynomial identities, remarkable identities* or *special products.* They are:

Polynomial or Remarkable identities	
Binomial squares	**Sum x Difference:**
$(a+b)^2 = a^2 + 2ab + b^2$	
$(a - b)^2 = a^2 - 2ab + b^2$	$(a + b) . (a - b) = a^2 - b^2$

These two figures may help you to memorize these identities.

$(a+b)^2 = a^2 + 2ab + b^2$

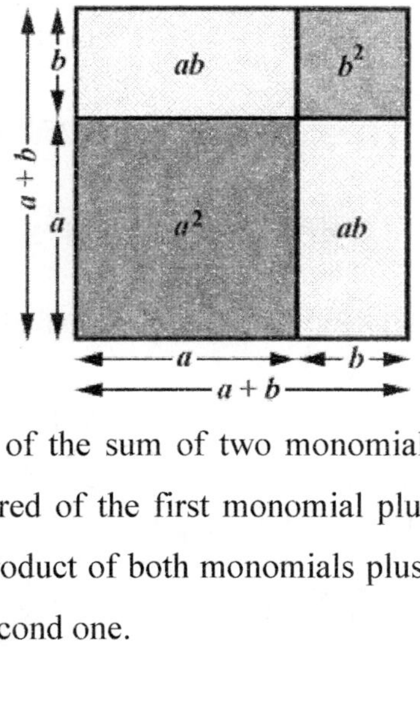

Squared of the sum of two monomials equals the squared of the first monomial plus double of the product of both monomials plus squared of the second one.

$(a - b)^2 = a^2 - 2ab + b^2$

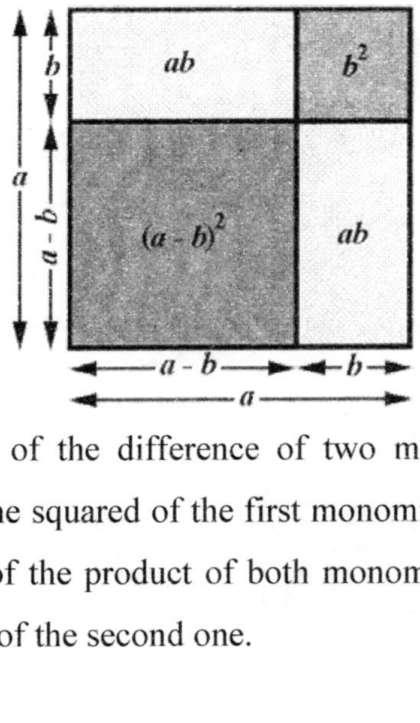

Squared of the difference of two monomials equals the squared of the first monomial minus double of the product of both monomials plus squared of the second one.

Example 8: Work out the following polynomial identities: a) $(x + 5)^2$ b) $(x - 6)^2$ c) $(x+2)\cdot(x-2)$

Solution:

a) $(x + 5)^2 = x^2 + 2\cdot x\cdot 5 + 5^2 = \boxed{x^2 + 10x + 25}$

b) $(3x - 4)^2 = (3x)^2 - 2\cdot 3x\cdot 4 + 4^2 = \boxed{9x^2 - 24x + 16}$ c) $(x + 2)\cdot(x - 2) = x^2 - 2^2 = \boxed{x^2 - 4}$.

Exercises

24. Work out the following polynomial identities:

a) $(x + 2)^2$ (Sol: $x^2 + 4x + 4$)

b) $(x - 3)^2$ (Sol: $x^2 - 6x + 9$)

c) $(x + 4)(x - 4)$ (Sol: $x^2 - 16$)

d) $(x + 3)^2$ (Sol: $x^2 + 6x + 9$)

e) $(x - 4)^2$ (Sol: $x^2 - 8x + 16$)

f) $(x + 5)(x - 5)$ (Sol: $x^2 - 25$)

g) $(a + 4)^2$ (Sol: $a^2 + 8a + 16$)

h) $(a - 2)^2$ (Sol: $a^2 - 4a + 4$)

i) $(a + 3)(a - 3)$ (Sol: $a^2 - 9$)

j) $(2x + 3)^2$ (Sol: $4x^2 + 12x + 9$)

k) $(3x - 2)^2$ (Sol: $9x^2 - 12x + 4$)

l) $(2x + 1)(2x - 1)$ (Sol: $4x^2 - 1$)

m) $(3x + 2)^2$ (Sol: $9x^2 + 12x + 4$)

n) $(2x - 5)^2$ (Sol: $4x^2 - 20x + 25$)

ñ) $(3x + 2)(3x - 2)$ (Sol: $9x^2 - 4$)

o) $(4b + 2)^2$ (Sol: $16b^2 + 16b + 4$)

p) $(5b - 3)^2$ (Sol: $25b^2 - 30b + 9$)

q) $(b + 1)(b - 1)$ (Sol: $b^2 - 1$)

r) $(4a + 5)^2$ (Sol: $16a^2 + 40a + 25$)

s) $(5a - 2)^2$ (Sol: $25a^2 - 20a + 4$)

t) $(5a + 2)(5a - 2)$ (Sol: $25a^2 - 4$)

u) $(4y + 1)^2$ (Sol: $16y^2 + 8y + 1$)

v) $(2y - 3)^2$ (Sol: $4y^2 - 12y + 9$)

w) $(2y + 3)(2y - 3)$ (Sol: $4y^2 - 9$)

x) $(3x + 4)^2$ (Sol: $9x^2 + 24x + 16$)

y) $(3x - 1)^2$ (Sol: $9x^2 - 6x + 1$)

z) $(3x^2 + 4)(3x^2 - 4)$ (Sol: $9x^4 - 16$)

a′) $(5b^3 + 1)^2$ (Sol: $25b^6 + 10b^3 + 1$)

b′) $(2x^2 - 4)^2$ (Sol: $4x^4 - 16x^2 + 16$)

c′) $(4x^3 + 3)(4x^3 - 3)$ (Sol: $16x^6 - 9$)

5. Factorization of a polynomial

If a, b, c, ... are the roots of a polynomial P(x), the **factorization** of P(x) is:

$P(x) = (x - a) \cdot (x - b) \cdot (x - c)......$

4. Steps to factorize a polynomial

Step 1. If possible, extract common factors:

Example: *a)* $x^3 + 6x^2 - 4x = x \cdot (x^2 + 6x - 4)$ *b)* $x^4 + 5x^3 - 3x^2 = x^2 \cdot (x^2 + 5x - 3)$

c) $3x^4 + 6x^3 - 3x^2 = 3x^2 \cdot (x^2 + 2x - 1)$

Step 2. Use either Ruffini´s Rule, special products or solve the quadratic equation to find out its roots.

Example: Factorize the polynomial $P(x) = x^4 + 6x^3 + 9x^2$.

Solution: - Step 1: extract common factors: $x^4 + 6x^3 + 9x^2 = x^2 \cdot (x^2 + 6x + 9)$

- Step 2: Special product: $x^2 + 6x + 9 = (x + 3)^2$

So. Factorization is: $x^4 + 6x^3 + 9x^2 = \boxed{x^2 \cdot (x + 3)^2}$

Exercises

25. Factorize the following polynomials:

a) $x^2 + 8x + 15$ b) $7x^2 - 21x - 280$ c) $3x^2 + 9x - 210$

26. By using the Reminder´s theorem, find out an integer root and, after that, factorize:

a) $2x^2 - 9x - 5$ b) $3x^2 - 2x - 5$ c) $4x^2 + 17x + 15$ d) $-x^2 + 17x - 72$

27. Extract a common factor and use the remarkable identities to factorize the polynomials:

a) $3x^3 - 12x$ b) $4x^3 - 24x^2 + 36x$ c) $45x^2 - 5x^4$

d) $x^4 + x^2 + 2x^3$ e) $x^6 - 16x^2$ f) $16x^4 - 9$

Exercises

28. Complete the factorization of the following expressions:

a) $(x^2 - 25)(x^2 - 6x + 9)$ b) $(x^2 - 7x)(x^2 - 13x + 40)$

29. Factorize and indicate the roots:

a) $x^3 + 2x^2 - x - 2$ b) $3x^3 - 15x^2 + 12x$

c) $x^3 - 9x^2 + 15x - 7$ d) $x^4 - 13x^2 + 36$

30. Factorize:

a) $x^3 + x^2$	b) $2x^4 + 4x^2$	c) $x^2 - 4$
d) $x^4 - 16$	e) $9 + 6x + x^2$	f) $x^4 - 10x^2 + 9$
g) $x^4 - 2x^2 - 3$	h) $x^4 - 4x^2$	i) $x^5 + 20x^3 + 100x$
j) $3x^5 - 18x^3 + 27x$	k) $2x^3 - 50x$	l) $2x^5 - 32x$
m) $2x^2 + x - 28$	n) $xy - 2x - 3y + 6$	ñ) $25x^2 - 1$
o) $36x^6 - 49$	p) $x^2 - 2x + 1$	q) $x^2 - 6x + 9$
r) $x^2 - 20x + 100$	s) $x^2 + 10x + 25$	t) $x^2 + 14x + 49$
u) $x^3 - 4x^2 + 4x$	v) $3x^7 - 27x$	w) $x^2 - 11x + 30$

Review exercises

1. Join each statement with its corresponding algebraic expression:

a) Squared of a number minus its double.

b) 80% of a number.

c) An odd number.

d) Two thirds of a number plus five units.

$\frac{2}{3}x + 5$; $x^2 - 2x$; $0{,}8x$; $2x + 1$

2. Express in algebraic language:

a) Product of two consecutive numbers.

b) Squared of a number plus its half.

c) Sum of a number and another one, 10 units higher.

3. Express in algebraic language perimeter and area of these rectangles:

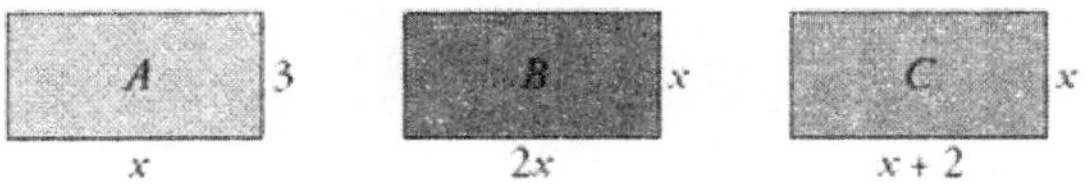

4. Translate into algebraic language using two unknowns:

a) The sum of the squared of two numbers.

b) The squared of the difference of two numbers.

c) Half of the product of two numbers.

d) Half of the sum of two numbers.

5. If x and y are the ages today of two brothers, express the following statements by using two unknowns:

a) The sum of their ages five years ago.

b) The product of their ages in 6 years.

c) Difference between the oldest and the youngest.

7. Indicate the degree of each monomial and which of them are like monomials:

a) $-5xy$ b) $(-7x)^3$ c) $8x$ d) $(xy)^2$ e) $\frac{2}{3}x^2y^2$ f) $\frac{4}{5}x^3$ g) $\frac{-3yx}{5}$ h) $\frac{1}{2}x^2$

6. Write the algebraic expression for the perimeter and area of these rectangles:

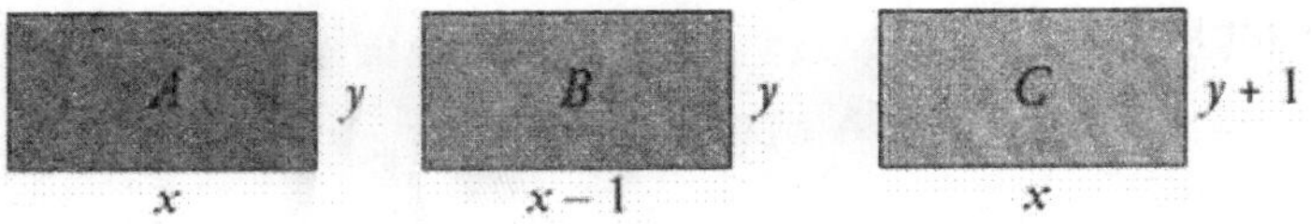

8. Simplify:

a) $6x^2 - 7x^2 + 3x^2$

b) $-6xy - 5xy + 10xy$

c) $\frac{1}{3}xy^2 - \frac{3}{5}xy^2 - \frac{7}{3}xy^2$

d) $\frac{2x^3}{3} + \frac{1}{5}x^3 - x^3$

9. Simplify:

a) $5x - x^2 + 7x^2 - 9x + 2$ b) $2x + 7y - 3x + y - x^2$ c) $x^2y^2 - 3x^2y - 5xy^2 + x^2y + xy^2$

10. Work out the following products of monomials:

a) $6x^2(-3x)$ b) $(2xy^2)(4x^2y)$ c) $\left(\frac{3}{4}x^3\right)\left(\frac{1}{2}x^3\right)$ d) $\left(\frac{1}{4}xy\right)\left(\frac{3xz}{2}\right)$

11. Simplify the following expressions.

a) $(2x^3 - 5x + 3) - (2x^3 - x^2 + 1)$ b) $5x - (3x + 8) - (2x^2 - 3x)$

What is the degree of the resulting polynmials?

12. Consider the folowing polynomials:

$A = 3x^3 - 5x^2 + x - 1$ $B = 2x^4 + x^3 - 2x + 4$ $C = -x^3 + 3x^2 - 7x$

Work out: $A + B$; $A - C$; $A - B + C$

Sol.: a) $2x^4 + 4x^3 - 5x^2 - x + 3$; b) $4x^3 - 8x^2 + 8x - 1$; c) $-2x^4 + x^3 - 2x^2 - 4x - 5$.

13. Work out, simplify and indicate the degree:

a) $x(x^2 - 5) - 3x^2(x + 2) - 7(x^2 + 1)$ b) $5x^2(-3x + 1) - x(2x - 3x^2) - 2\cdot 3x$

c) $\frac{1}{3}x^2\left(-\frac{3}{2}x^2 + 6x - 9\right)$

Sol.: a) $-2x^3 - 13x^2 - 5x - 7$ (Degree 3); b) $-12x^3 + 3x^2 - 6x$ (Degree 3); c) $-\frac{1}{2}x^4 + 2x^3 - 3x^2$.

(Degree 4)

14. Work out and simplify:

a) $(2x^2 + 3)(x - 1) - x(x - 2)$ b) $(x + 4)(2x^2 + 3x - 5) - 3x(-x + 1)$

c) $(x^2 - 5x + 3)(x^2 - x) - x(x^3 - 3)$ d) $\left(\frac{1}{2}x^2 + \frac{5}{3}x + \frac{1}{6}\right)(6x - 12)$

Sol.: a) $2x^3 - 3x^2 + 5x - 3$; b) $2x^3 + 14x^2 + 4x - 20$; c) $-6x^3 + 8x^2$; d) $3x^3 + 4x^2 - 19x - 2$

15. Factorize:

a) $12x^3 - 8x^2 - 4x$ b) $-3x^3 + x - x^2$ c) $2xy^2 - 4x^2y + x^2y^2$ d) $\frac{2}{3}x^2 + \frac{1}{3}x^3 - \frac{5}{3}x$

Sol.: a) $4x(3x^2 - 2x - 1)$; b) $x(-3x^2 + 1 - x)$; c) $xy(2y - 4x + xy)$; d) $\frac{1}{3}x(2x + x^2 - 5)$

16. Factorize, as in the example:

Example: $3x(x+1) - x^2(x+1) + (x+1)(x^2-2) = (x+1)[3x - x^2 + x^2 - 2] = (x+1)(3x-2)$

a) $2x(x-2) + x^2(x-2) - 3(x-2)$ b) $x^2(x+1) - x^2(x+2) + 2x^2(x-3)$

c) $3x^2(x+3) - 6x(x+3)$ **Sol.:** a) $(x-2)(2x+x^2-3)$; b) $x^2(2x-7)$; c) $x(x+3)(3x-6)$.

17. Expand:

a) $(x+6)^2$ b) $(7-x)^2$ c) $(3x-2)^2$ d) $\left(x+\frac{1}{2}\right)^2$

e) $(x-2y)^2$ f) $\left(\frac{2}{5}x-\frac{1}{3}y\right)^2$

Sol.: a) $x^2+36+12x$; b) $49+x^2-14x$; c) $9x^2+4-12x$; d) $x^2+\frac{1}{4}+x$;

e) x^2+4y^2-4xy; f) $\frac{4x^2}{25}+\frac{y^2}{9}-\frac{4xy}{15}$

18. Work out:

a) $(x+7)(x-7)$ b) $(3+x)(3-x)$ c) $(3+4x)(3-4x)$

d) $(x^2+1)(x^2-1)$ e) $\left(\frac{1}{2}x-1\right)\left(\frac{1}{2}x+1\right)$ f) $\left(1+\frac{1}{x}\right)\left(1-\frac{1}{x}\right)$

Sol.: a) x^2-49; b) $9-x^2$; c) $9-16x^2$; d) x^4-1; e) $\frac{x^2}{4}-1$; f) $1-\frac{1}{x^2}$

19. Simplify as much as possible:

a) $(x+3)(x-3)-(x+3)^2$ b) $(2x+3)^2-(2x-3)^2-9$

c) $3x(x+1)^2-(2x+1)(2x-1)$ d) $(x^2+2)(x^2-2)-(x^2-1)^2$

Sol.: a) $6x-18$; b) (-9); c) $3x^3+2x^2+3x+1$; d) $2x^2-5$.

20. Express as a difference of squares:

a) $(2x+7)(2x-7)$ b) $(4x-1)(4x+1)$ c) $(x^2+x)(x^2-x)$

d) $(1-5x)(1+5x)$ **Sol.:** a) $4x^2-49$; b) $16x^2-1$; c) x^4-x^2; d) $1-25x^2$.

21. Complete the gaps:

a) $x^2 + \ldots + 4x$ b) $x^2 + \ldots - 10x$ c) $x^2 + 9 + \ldots$ d) $x^2 + 16 - \ldots$

22. Express as a squared of a sum or a squared of a difference, as in the example:

Example: $x^2 + 25 + 10x = x^2 + 5^2 + 2 \cdot 5x = (x + 5)^2$

a) $x^2 + 49 - 14x$ b) $x^2 + 1 - 2x$ c) $4x^2 + 1 + 4x$ d) $x^2 + 12x + 36$

Sol.: a) $(x-7)^2$; b) $(x-1)^2$; c) $(2x+1)^2$; d) $(x+6)^2$.

23. Simplify the following algebraic fractions:

a) $\frac{9x}{12x^2}$ b) $\frac{x(x+1)}{5(x+1)}$ c) $\frac{x^2(x+2)}{2x^3}$ **Sol.:** a) $\frac{3}{4x}$; b) $\frac{x}{5}$; c) $\frac{(x+2)}{2x}$

24. Simplify:

a) $\frac{x^2-4x}{x^2}$ b) $\frac{3x}{x^2+2x}$ c) $\frac{3x+3}{(x+1)^2}$ d) $\frac{2x^2+4x}{x^3+2x^2}$

e) $\frac{8x^3-4x^2}{(2x-1)^2}$ f) $\frac{5x^3+5x}{x^4+x^2}$ **Sol.:** a) $\frac{x-4}{x}$; b) $\frac{3}{x+2}$; c) $\frac{3}{(x+1)}$; d) $\frac{2}{x+2}$; e) $\frac{4x^2}{2x-1}$; f) $\frac{5}{x}$

25. Express as an algebraic expression the area of the coloured zone:

Sol.: 4x + 4y – 16.

26. Express as an algebraic expression the area and the longest diagonal of this trapezium:

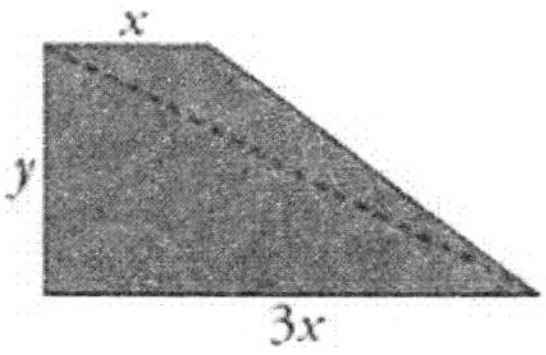

Sol.: $A = 2xy$; $d = \sqrt{y^2 + 9x^2}$.

27. Simplify the followig expressions:

a) $6\left(\frac{5x-4}{6}+\frac{2x-3}{2}-\frac{x-1}{3}\right)$ b) $12\left(\frac{x+6}{3}-\frac{x+1}{2}+\frac{3x-1}{4}\right)$

c) $4\left((x-2)^2-\frac{3}{4}x^2-4\right)$

d) $30\left(\frac{x(x-2)}{15}-\frac{(x+1)^2}{6}+\frac{1}{2}\right)$ **Sol.:** a) 9x – 12; b) 7x + 15; c) $x^2 - 8x$; d) $-3x^2 - 14x + 10$.

28. Express as the squared of a sum, the squared of a difference or a difference of squares:

a) x^2+9-6x b) $4x^2+1+4x$ c) $4x^2-9$

d) $9x^2-12x+4$ e) $16x^2-1$ f) $16x^2+40x+25$

Sol.: a) $(x-3)^2$; b) $(2x+1)^2$; c) $(2x+3)(2x-3)$; d) $(3x-2)^2$; e) $(4x+1)(4x-1)$; f) $(4x+5)^2$.

29. Transform into a product, as in the example:

Example: $x^3+2x^2+x=x(x^2+2x+1)=x(x+1)^2$

a) x^3-4x b) $4x^3-4x^2+x$ c) x^4-x^2 d) $3x^4-24x^3+48x^2$

Sol.: a) $x(x+2)(x-2)$; b) $x(2x-1)^2$; c) $x^2(x+1)(x-1)$; d) $3x^2(x-4)^2$.

30. Simplify. First of all, transform numerator and denominator into product:

a) $\dfrac{2x+4}{3x^2+6x}$ b) $\dfrac{x+1}{x^2-1}$ c) $\dfrac{x-2}{x^2+4-4x}$ d) $\dfrac{x^2-3x}{x^2-9}$ e) $\dfrac{x^2-4}{x^2+4x+4}$ f) $\dfrac{x^3+2x^2+x}{3x+3}$

Sol.: a) $\dfrac{2}{3x}$; b) $\dfrac{1}{x-1}$; c) $\dfrac{1}{x-2}$; d) $\dfrac{x}{x+3}$; e) $\dfrac{x-2}{x+2}$; f) $\dfrac{x(x+1)}{3}$.

31. Work out the following divisions:

a) $(8x^4+3x^3+2x-2):(4x^2+x-3)$ (Sol: $Q(x)=2x^2+x/4+23/16$; $R(x)=21x/16+37/16$)

b) $(2x^5-x^3+3x-9):(2x^2-x+2)$ (Sol: $Q(x)=x^3+x^2/2-5x/4-9/8$; $R(x)=35x/8-27/4$)

c) $(6x^3-3x^2+2x-5):(3x-2)$ (Sol: $Q(x)=2x^2+x/3+8/9$; $R(x)=-29/9$)

e) $(4x^4-x^3+x+5):(2x^2-x+3)$ (Sol: $Q(x)=2x^2+x/2-11/4$; $R(x)=-13x/4+53/4$)

f) $(6x^4+3x^3-5x^2+x-8):(3x^2-5x+2)$ (Sol: $Q(x)=2x^2+13x/3+38/9$; $R(x)=121x/9-148/9$)

g) $(8x^4-3x^2+7x-5):(4x^2-3x+2)$ (Sol: $Q(x)=2x^2+3x/2-5/8$; $R(x)=17x/8-15/4$)

h) $(6x^5+5x^4+31x^2+2):(2x^2+2)$ (Sol: $Q(x)=3x^3+5x^2/2-3x+13$; $R(x)=6x-24$)

i) $(3x^5-6x^4-x^3+10x^2-8x+2):(3x^2-6x+1)$ (Sol: $Q(x)=x^3-2x/3+2$; $R(x)=14x/3$)

j) $(6x^4-x^3+2x^2-x-1):(3x^2+2)$ (Sol: $Q(x)=2x^2-x/3-2/3$; $R(x)=-x/3+1/3$)

32. Work out the following divisions using the Ruffini´s rule:

a) $(x^4+x^3-x^2+x-1):(x+2)$ (Sol: $Q(x)=x^3-x^2+x-1$; $R=1$)

b) $(x^3-7x^2/2-10x/3-70):(x-6)$ (Sol: $Q(x)=x^2+5x/2+35/3$; División exacta)

c) $(2x^3+3x^2-1):(x-1/2)$ (Sol: $Q(x)=2x^2+4x+2$; División exacta)

d) $(3x^3+2x^2+2x-1):(x-1/3)$ (Sol: $Q(x)=3x^2+3x+3$; División exacta)

e) $(ax^3-3a^2x^2+2a^3x+1):(x-a)$ (Sol: $Q(x)=ax^2-2a^2x$; $R=1$)

33. Find the value of *k* so that -2 is a root of polynomial x^3-5x^2-7x+k. **Sol.:** $k=14$.

34. What is the result of multiplying an algebraic fraction by its inverse? Check it with the fraction $\frac{x}{x+2}$. **Sol.:** 1.

35. a) Simplify this expression: $a^2 - (a+1)(a-1)$.

b) Answer in less than 2 seconds: What is the value of $7500^2 - 7501 \cdot 7499$? **Sol.:** a) 1; b) 1.

36. a) Simplify the expression $(a+1)^2 - (a-1)^2$.

b) Answer in less than 2 seconds: What is the value of $2501^2 - 2499^2$? **Sol.:** a) 4a; b) 10,000.

37. Factorize (Two first are made for you):

a) $x^2 - 25 = \boxed{(x+5)\cdot(x-5)}$

b) $x^2 - 18x + 81 = \boxed{(x-9)^2}$

c) $x^2 - 9$

d) $x^2 - 20x + 100$

e) $x^2 - 81$

f) $x^2 - 14x + 49$

g) $x^2 - 100$

h) $x^2 + 22x + 121$

i) $x^2 - 49$

j) $x^2 + 18x + 81$

k) $9x^2 - 16$

l) $16x^2 - 9$

m) $x^2 - 36$

n) $25x^2 - 120x + 144$

ñ) $49x^2 + 84x + 36$

o) $x^2 - 16x + 64$

p) $81x^2 - 180x + 100$

q) $49x^2 - 64$

r) $x^2 - 1$

s) $25x^2 - 1$

t) $4x^2 - 12x + 9$

u) $x^2 - 6x + 9$

v) $x^2 + 10x + 25$

w) $x^2 - 121$

38. Reduce the following algebraic expressions to their simplest form:

a) $\frac{6x-4}{9x^3-6x^2}$ b) $\frac{5x^2+10x}{x+2}$ c) $\frac{x^3+x^2}{2x^3-3x^2}$ d) $\frac{3x^3-x^2}{x^3+2x^2}$ e) $\frac{x^2+2x+1}{x^2-1}$ f) $\frac{x^2-4}{x^2-4x+4}$

g) $\frac{2x^2-8}{x+2}$ h) $\frac{2x+1}{4x^2+4x+1}$ i) $\frac{2x^4-2x^3}{4x^4-4x^2}$ j) $\frac{3x^4-9x^2}{x^2-3}$

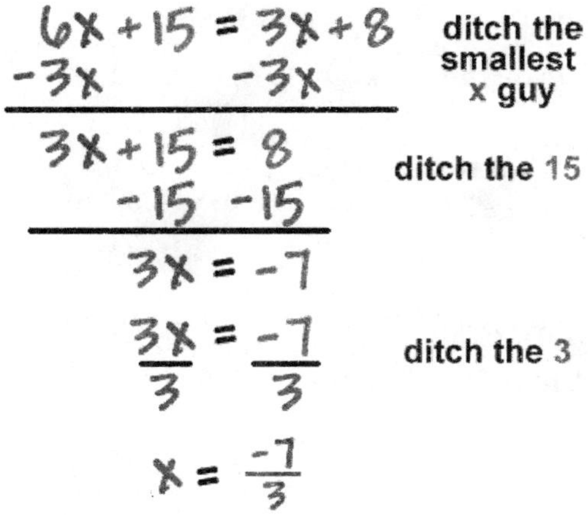

Unit 5.- Equations

1. First degree equations with one unknown

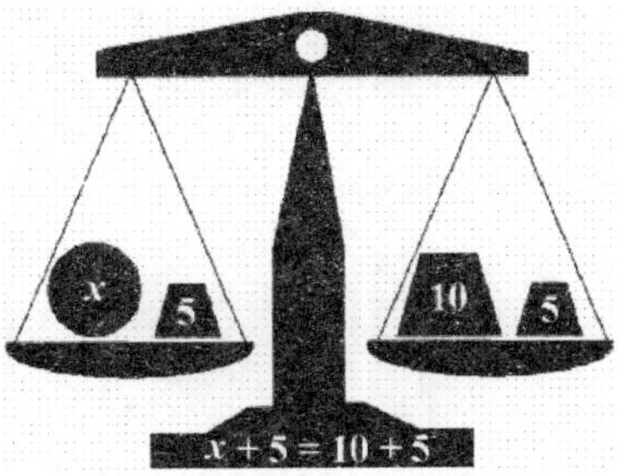

As you can see in the figure, plates of the balance are equilibrated. So, there is an equality relationship between the masses of both plates.

We obtain an *equation*. Concretely, this one is named *first degree equation*. This means that the unknown has an exponent 1.

The following are also examples of first degree equations:

$$3x = 12 \qquad 3m + 7 = 2m - 6$$

$$6t + 7 - 2t = 3t + 1$$

Solutions or ***roots*** of an equation are the values that, when substituted by the unknown, make the equality to verify.

Solving an equation is finding out the solutions or roots if the equation.

Depending on the solutions, an equation can be:

· ***Compatible*** if equation has solutions. If the number of solutions is finite, equation is named ***determined compatible***, and if the number of solutions is infinite, equation is named ***non-determined compatible.***

· ***Incompatible*** or ***impossible*** when equation has not solution.

Exercises

1. Check if the given values are solutions of the following equations:

a) $2x + 3 = 5x - 1 \rightarrow x = 3$

b) $\frac{x}{2} + 1 = x - 1 \rightarrow x = 4$

c) $3(x - 2) = 2x + 4 \rightarrow x = 10$

d) $\frac{5x-2}{10} + 3 = 2x + 1 \rightarrow x = (-1)$

2. Write two equations having as solution x = (-2).

2. Equivalent equations. Equivalence transformations

Two or more equations are said to be ***equivalent*** if they have the same solution.

When we want to solve an equation, we must transform it into a more simple equation, equivalent to the original. This is possible by using the ***Equivalence Transformations:***

Transformation		Practical rule
Adding or subtracting the same expression in both members of the equality.	→	What it is adding in a member, passes subtracting to the other member, and vice versa.
Multiplying or dividing both members of equality by the same number.	→	What it is multiplying in a member, passes dividing to the other member, and vice versa.

Example: By applying above instructions, solve these equations:

a) $2x - 5 = x + 7$ b) $3x - 7 = 41$

<u>Solution:</u>

a) $2x - 5 = x + 7$

Add 5 to both members: $2x = x + 12$

Subtract x to both members $x = 12$

Solution is: $\boxed{x = 12}$

b) $3x - 7 = 41$

Add 7 to both members: $3x = 48$

Divide both members by 3: $x = 16$

Solution is: $\boxed{x = 16}$

Exercises

3. Solve these equations:

a) $2x - 7 = 3x - 8$ b) $2x - 4 = 3x + 2$ c) $2x - 14 = -4x + 4$

d) $18 + 3x = 2x - 12$ e) $10 + 2x = 6x + 2$ f) $9x - 9 = 6x + 18$

g) $8 - 4 + x = 9$ h) $12 - x + 3 = 6$ i) $7 - 2x + 3 = 2$

4. Apply equivalence transformations to solve the following equations:

a) $2x - 5 = x + 41 - 5$ b) $3x + 1 = x + 7$ c) $3x + 23 = 2x + 59$

d) $2x + \frac{5x}{3} = 66$ e) $5x + 8 = 8x + 2$ f) $9 + 9x = 117 - 3x$

g) $21 - 7x = 41x - 123$ h) $500 - 24x = - 4 - 3x$

2.1. Steps to solve a linear equation

1.- Expand all **parentheses** in the equation.

2.- Eliminate **denominators** by multiplying every term by the LCM of the denominators.

3.- Move terms: Terms containing "x" must be together in one member, and non containing "x" terms, in the other one.

4.- Simplify **like terms** at both members, having an equation as **ax = b**.

5.- Calculate the value of the **unknown**.

Example: Solve the equation $\frac{x}{3}-\frac{3(x-2)}{5}=2\left(\frac{x-1}{3}\right)$

Solution:

Step 1: Parentheses: $\frac{x}{3}-\frac{3x-6}{5}=\frac{2x-2}{3}$

Step 2: Denominators: LCM (3,5, 3) = 15. We multiply each term by 15:

$5x - 3(3x - 6) = 5(2x - 2)$ We expand new parentheses: $5x - 9x + 18 = 10x - 10$

Step 3: Move terms: $5x - 9x - 10x = -10 - 18$

Step 4: Simplify like terms: $-14x = -28$

Step 5: and, finally, $x=\frac{-28}{-14}=\boxed{2}$.

Exercises

5. Solve these equations:

a) 10 + x + 14 = 30 + 5

b) 18 + 2x – 8 = x – 25

c) 8x – 6 = x + 8 + 6x

d) 4x – 12 + x = 4x – 1

e) 3x = - 27

f) 5x + 4 = 20 + 2x

6. Solve:

a) 3 (x + 6) = -2 (5 – x)

b) x + 9 = 2 (x – 6)

c) 2x + 3= 4x + 6 (x – 4) – 2

d) 1 + 4(x – 2) = - 3x + 5(x + 1)

e) 2(x + 6) – 7x = 3x – 5x + 8

f) 2(x + 3) – 6(x + 5) = 3x + 4

g) –2x + 3(x – 1) = - 12 + 5(2 – x)

h) 5(x – 1) – 6x = 3(x – 3)

Exercises

7. Solve:

a) $\frac{10x-55}{2}=10x-\frac{95-10x}{2}$

b) $x+5=\frac{x+3}{3}$

c) $\frac{-4x+12}{4}=x-5$

d) $\frac{3+x}{2}=4$

e) $\frac{x}{2}+21=\frac{4x}{3}+24$

f) $2(x+3)-7=3\left(\frac{1}{2}-2x\right)$

g) $\frac{5-9x}{8}+\frac{2x+3}{4}-\frac{143}{6}=2x$

h) $\frac{5x+7}{2}-\frac{3x+9}{4}=\frac{2x+4}{3}+5$

i) $1-\frac{x-5}{4}-\frac{x-3}{10}+\frac{x+3}{8}=0$

8. Solve these equations:

a) $\frac{3(x+2)}{2}+\frac{x-1}{5}=\frac{2(x+1)}{5}+\frac{37}{10}$

b) $\frac{2x-3}{2}-\frac{x+3}{4}=-4-\frac{x-1}{2}$

c) $\frac{1+12x}{4}+\frac{x-4}{2}=\frac{3(x+1)-(1-x)}{8}$

d) $\frac{3x-2}{6}-\frac{4x+1}{10}=-\frac{2}{15}-\frac{2(x-3)}{4}$

e) $\frac{2x-3}{6}-\frac{3(x-1)}{4}-\frac{2(3-x)}{6}+\frac{5}{8}=0$

f) $\frac{2}{3}(x+3)-\frac{1}{2}(x+1)=1-\frac{3}{4}(x+3)$

9. Solve these equations:

a) $3(x+6)=-2(5-x)$

b) $1+4(x-2)=-3x+5(x+1)$

c) $2(x+3)-6(x+5)=3x+4$

d) $3(5x+9)-3(x-7)=11(x-2)$

e) $\frac{10x-55}{2}=10x-\frac{95-10x}{2}$

f) $\frac{x+4}{5}-\frac{x+3}{4}=1-\frac{x+1}{2}$

g) $\frac{3x-7}{12}=\frac{2x-3}{6}-\frac{x-1}{8}$

h) $2+\frac{3x-1}{15}+\frac{x-4}{5}=\frac{x+4}{3}$

2.2. Number of solutions of a linear equation

As we have seen, all linear equations can be transformed into an equivalent one having the form **ax = b**.

Depending of the values of *a* and *b*, the equation is going to a have a different number of solutions. We are going to see it with some examples:

- $\mathbf{180 - 3x = x + 12} \rightarrow -3x - x = 12 - 180 \rightarrow -4x = -168 \rightarrow x = 42.$

  This equation has an only solution.

- $\mathbf{7(x + 2) - 5x = 2x + 14} \rightarrow 7x + 14 - 5x = 2x + 14 \rightarrow 7x - 5x - 2x = 14 - 14 \rightarrow 0x = 0.$

  This equation has infinite solutions. All real numbers can be substituted by "x".

- $\mathbf{2(x + 2) = 2x + 7} \rightarrow 2x + 4 = 2x + 7 \rightarrow 2x - 2x = 7 - 4 \rightarrow 0x = 3.$

  This equation has not solution. We cannot substitute any number by "x".

Remember:

<u>**Number of solutions of a linear equation**</u>

1°. Transform the equation into another equivalent one having the form ***ax = b.***

2°. Pay attention to:

If $\boldsymbol{a \neq 0, ax = b}$, equation has an ***only solution*:** $x = \boldsymbol{b/a}$, so, it is a ***determined compatible*** equation.

If $\boldsymbol{a = b = 0, 0x = 0}$, equation has **infinite solutions** (all real numbers are solution), so, it is an ***undetermined compatible*** equation.

If $\boldsymbol{a = 0}$ and $\boldsymbol{b \neq 0, 0x = b}$, equation has ***no solution***, and it is an ***incompatible*** equation.

Exercises

10. Solve and classify these equations:

a) $3x + 100 = 5(200 - 3x)$

b) $5(20 - x) = 4(2x - 1)$

c) $-10x + 8 = 2(4 - 5x)$

d) $9(x - 2) - 3(x - 4) = 3(16 - 7x)$

e) $6 - (3x + 1) = 5 - 3(x - 2)$

f) $4 - 2(x - 1) = 3(2 - x) - 10$

g) $\dfrac{5x-1}{7} - \dfrac{7x+19}{20} = \dfrac{x+11}{14} + \dfrac{3x-4}{5}$

h) $\dfrac{x}{3} + 3 - \dfrac{2x}{5} = \dfrac{1}{5}\left(-\dfrac{x}{3} + 15\right)$

2.3. Problems with linear equations

Exercises

11. A pride of lions is made up of 13 lions, and there are 3 females more than males. How many lions and how many lionesses are there in the pride?

12. There are 31 people in a bar. How many men and women are there if we know that there are more men than women?

13. In a cow farm, the number of horns and legs add up to 30. How many cows are there in the farm?

14. When a number x is multiplied by 5, it gives the same result as when 48 is added to twice the number. Write down an equation for x and find its solution.

15. When 8 is added to a certain number, the result is 3 times as large as when 2 is subtracted from the number. Find the original number.

16. 60 people, between men, women and children, take part in a touristic trip. Children are double than men, and women number is 2/3 of the number of men and children together. How many men, women and children are there?

17. Find three consecutive multiples of 3 whose sum is 702.

Exercises

18. Today, Mary´s age is double than Ann´s age. In 12 years, Mary´s age will be three times Ann´s age. How old are Mary and Ann today?

19. If we add 12 to certain number, calculate the double of the resulting quantity, and finally, subtract 24 units to the resulting number, we will obtain the double of the original number. What was the original number?

20. In a car park, there are 452 vehicles, between cars and motorbikes. Find out the number of vehicles of each kind, knowing there 1,744 wheels touching the floor.

21. In my instituto, we are 300 students (boys and girls). Today, 155 of them have had a cultural trip. Knowing they have taken part in the trip 60% of boys in the instituto and 40% of girls, how many boys and girls are there in the instituto?

22. A group of friends won a cash prize which gave them 28 € each. If there had been 5 more friends in the group they would each have received 7 € less. How many friends won the prize?

23. Janet is three times as old as her daughter, Mary. Five years ago, Janet was four times as old as Mary. How old is Janet now? How old will Mary be in 7 years?

3. Quadratic equations

3.1. Quadratic equations

It is named ***quadratic equation*** an equation that can be transformed into another equivalent equation having this form:

$$ax^2 + bx + c = 0, \quad \text{with } a \neq 0$$

Solutions of a quadratic equation are the values of x that being substituted in it, make the equation to verify.

Quadratic equations are solved by using this expression: $x = \dfrac{-b \pm \sqrt{b^2 - (4 \cdot a \cdot c)}}{2 \cdot a}$

Example: Solve: $x^2 - 3x + 2 = 0$

Solution:

$$\left.\begin{array}{l} a=1 \\ b=(-3) \\ c=2 \end{array}\right\} x = \frac{-(-3) \pm \sqrt{(-3)^2 - (4 \cdot 1 \cdot 2)}}{2 \cdot 1} = \frac{+3 \pm \sqrt{9-8}}{2} = \frac{+3 \pm 1}{2} \rightarrow \begin{cases} x_1 = \frac{+3+1}{2} = \frac{4}{2} = 2 \\ x_2 = \frac{+3-1}{2} = \frac{2}{2} = 1 \end{cases}$$

So, solutions are $\boxed{2}$ and $\boxed{1}$.

Exercises

24. Find the values of a, b and c in the following quadratic equations:

a) $3x^2 - 8x + 9 = 0$ b) $4x^2 - 3x + 2x^2 = 0$ c) $2x - 6 = 3x^2$

d) $3x + 2 = x^2 - 5x$ e) $x^2 - 6 = 2$ f) $0 = 3x^2 - 5x + 6$

25. Solve these quadratic equations:

a) $x^2 - 6x + 8 = 0$ b) $x^2 - 8x + 7 = 0$ c) $4x^2 + 24x + 20 = 0$

d) $x^2 + 2x - 8 = 0$ e) $x^2 - 4x - 21 = 0$ f) $3x^2 + 6x - 12 = 0$

26. Solve these quadratic equations:

a) $2x^2 + 4x - 16 = 0$ b) $3x^2 - 6x - 9 = 0$ c) $x^2 + 9x + 20 = 0$

d) $x^2 + 4x + 4 = 16$ e) $x^2 - 6x + 9 = 25$ f) $4x^2 + 4x + 1 = 9$

3.2. Incomplete quadratic equations

In a quadratic equation $\boldsymbol{ax^2 + bx + c = 0}$, if parameters b or c, are zero, it is named an ***incomplete quadratic equation.***

Incomplete quadratic equations can be solved by using the general expression we have just studied, or by a shorter method:

· If $\boldsymbol{b = 0}$, equation becomes $\boldsymbol{ax^2 + c = 0}$. It is quickly solved.

· If $\boldsymbol{c = 0}$, equation becomes $\boldsymbol{ax^2 + bx = 0}$. It is solved by factorizing. **x(ax + b) = 0.**

Example: Solve: a) $3x^2 = 0$ b) $2x^2 - 8 = 0$ *c)* $x^2 - 12x = 0$

Solution:

a) $3x^2 = 0$ Dividing by 3: $x^2 = 0 \rightarrow x = \sqrt{0} = \boxed{0}$.

b) $2x^2 - 8 = 0$ Adding 8: $2x^2 = 8 \rightarrow$ Dividing by 2: $x^2 = 4 \rightarrow x = \sqrt{4} = \boxed{+2 \text{ and } (-2)}$.

c) $x^2 - 12x = 0$ Factorizing: $x(x - 12) = 0 \rightarrow \begin{cases} x = 0 \\ x - 12 = 0 \rightarrow x = 12 \end{cases}$

For this last step, you must remember if you have a·b = 0, then, there are 2 possible solutions: $\begin{cases} a = 0 \\ b = 0 \end{cases}$

Exercises

27. Solve these incomplete quadratic equations:

a) $3x^2 - 12 = 0$ b) $5x^2 - 45 = 0$ c) $x^2 - 16 = 0$

d) $2x^2 - 6x = 0$ e) $6x^2 - 18x = 0$

28. Solve these incomplete quadratic equations:

a) $5x^2 - 20 = 0$ b) $3x^2 + 5x = 0$ c) $3x^2 - 75 = 0$ d) $5x^2 + 6x = 0$

e) $3x^2 = 0$ f) $x^2 - 16 = 0$ g) $4x^2 + 7x = 0$ h) $5x^2 - 15x = 0$

i) $2x^2 = 7x$ j) $3x^2 = 18x$

29. Solve these quadratic equations:

a) $x^2 - 1 = 0$ b) $x^2 - 16 = 0$ c) $x(x + 1) = 0$ d) $x(x - 3) = 0$

e) $(x - 1)^2 = 0$ f) $x^2 - 4 = 0$ g) $x^2 - 81 = 0$ h) $x(x - 1) = 0$

i) $x(x + 5) = 0$ j) $(x - 3)^2 = 0$

30. Solve these incomplete quadratic equations, <u>without operating</u>:

a) $(3x + 5)\cdot(2x - 1) = 0$ b) $(6x - 3)\cdot(2x - 1) = 0$ c) $(2x - 4)\cdot(x + 6) = 0$

d) $(5x - 8)\cdot(x - 4) = 0$ e) $(x-4)(x-6) = 0$ f) $(x+2)(x-3) = 0$

g) $x(x+1)(x-5) = 0$ h) $(3x+1)(2x-3) = 0$

Exercises

31. Solve these quadratic equations:

a) $x^2+2x=0$	(0 y -2)	b) $2x^2+5x-3=0$	(1/2 y -3)
c) $x^2-24=1$	(5 y -5)	d) $1-4x^2=-8$	(3/2 y -3/2)
e) $x^2-6x=0$	(0 y 6)	f) $3x^2-39x=0$	(0 y 13)
g) $3x^2+15x+18=0$	(-2 y -3)	h) $-x^2+4x-7=0$	(∄)
i) $3x^2-48=0$	(4 y -4)	j) $x^2-3x+2=0$	(1 y 2)
k) $x^2-3x+1=0$	(2,6 y 0,3)	l) $3x^2-47=1$	(4 y -4)

32. Solve these quadratic equations:

a) $3x^2-16=2x^2$	(4 y -4)	b) $-x^2+9x=0$	(0 y 9)
c) $x^2-7x-18=0$	(-2 y 9)	d) $7x^2+21x-28=0$	(1 y -4)
e) $x^2-10x+9=0$	(9 y 1)	f) $x^2-26x+25=0$	(25 y 1)
g) $4x^2-37x+9=0$	(9 y ¼)	h) $4x^2-17x+4=0$	(4 y ¼)
i) $x^2-25x+144=0$	(9 y 16)	j) $4x^2-9x+2=0$	(2 y ¼)
k) $2x^2-5x+3=0$	(1,3/2)		

3.3. Number of solutions of a quadratic equation

As you have already seen by solving all the previous quadratic equations, these equations can have:

- 2 solutions. This is the most general situation.
- 1 solution.
- No solution.

Discriminant

In the expression $x = \dfrac{-b \pm \sqrt{b^2 - (4 \cdot a \cdot c)}}{2 \cdot a}$, quantity into the square root is named ***discriminant*** (Δ).

So, discriminant is Δ = $\mathbf{b^2 - (4 \cdot a \cdot c)}$

Value of the discriminant allows to know the number of solutions of a quadratic equation, without having to calculate them:

- If **D > 0** equation has **two solutions.**
- If **D = 0** equation has **one solution** (named *double solution*).
- If **D < 0** equation has **not solution.**

Example: Find the value of k so that equation $x^2 + kx + 36 = 0$ has two identical solutions.

Solution:

If we want the equation to have only one (double) solution, we need D = 0.

Substituting: $b^2 - 4ac = k^2 - 4 \cdot 1 \cdot 36 = k^2 - 144 = 0 \rightarrow k^2 = 144 \rightarrow k = \sqrt{144} = \pm 12$

So, equations are: $x^2 + 12x + 36 = 0$ and $x^2 - 12x + 36 = 0$.

Exercises

33. Solve the following quadratic equations:

a) $3x^2 - 5x + 2 = 0$ b) $16x^2 + 8x + 1 = 0$ c) $2x(x + 3) = 3(x - 19)$

d) $2x^2 + 4x + 4 = 0$ e) $x^2 - 7x = 18$ f) $x^2 + 9 = 10x$

34. Solve the following quadratic equations:

a) $2x(3x - 4) + 2 = (1 - 3x)(x + 1)$ b) $4x^2 - 17x = -4$

c) $4x^2 - 37x + 9 = 0$ d) $x^2 + x + 1 = 0$

Exercises

35. In equation x^2 - $5x + c = 0$, one solution is x = 3. a) What is the value of coefficient c? b) What is the other solution?

36. Find out the value of m in equation x^2 - $6x + m = 0$, so that it has two identical solutions. Find also the solutions.

37. Calculate the discriminant of each equation and, without solving it, indicate their number of solutions:

a) $5x^2 - 3x + 1 = 0$ b) $x^2 - 4x + 4 = 0$ c) $3x^2 - 6x - 1 = 0$

d) $5x^2 + 3x + 1 = 0$

38. Determine for what values of **m** equation $2x^2 - 5x + m = 0$:

a) Has two different solutions. b) Has one solution. c) Has not solution.

39. Decide for what values of **b** equation $x^2 - bx + 25 = 0$:

a) Has two different solutions. b) Has one solution. c) Has not solution.

3.4. Problems with quadratic equations

Exercises

40. The area of a parallelogram is 50 cm^2. If the base is twice its height, calculate the height.

41. The width of a rectangular plot of land is 5 m less than its length. If the area of the plot is 104 m^2, find the dimensions of the plot.

42. A circle has an area of 154 cm^2. Find its radius.

43. In a triangle, its base is 3 cm less than its height. If its area is 14 cm^2, find its height.

Exercises	**44.** The area of a rectangle is 52 cm^2. Find the length and the width of this rectangle if their difference is 14 cm. **45.** When 2 is added to a certain number, the result is the same as dividing 8 by the number. Solve this problem finding the number/s. **40.** The area of a parallelogram is 50 cm^2. If the base is twice its height, calculate the height.

Review exercises

Linear equations

1. Solve the following equations and check solutions:

a) $3x - 2(x + 3) = x - 3(x + 1)$

b) $4 + x - 4(1 - x) + 5(2 + x) = 0$

c) $2x + 7 - 2(x - 1) = 3(x + 3)$

d) $4(2x - 7) - 3(3x + 1) = 2 - (7 - x)$

2. Check if these equations are equivalent:

$2(x - 1) + x + 1 = 2x + 1$ $\qquad$ $2x - 1 - (x - 1) = 2(3x - 5)$

3. Solve the following equations:

a) $\frac{x}{5} + \frac{3x}{4} - \frac{5x}{6} = 15$

b) $4 \cdot (x-3) - 7 \cdot (x-4) = 6 - x$

c) $5x + 10 = 8x + 4$

d) $11 + 9x = 119 - 3x$

e) $508 - 23x = 4 - 2x$

f) $5 \cdot (20 - x) = 4 \cdot (2x - 1)$

g) $4 \cdot (x-3) - 7 \cdot (x-4) = 6 - x$

h) $\frac{x-1}{1} - \frac{x-2}{2} + \frac{x-3}{3} = 0$

i) $\frac{3x-11}{20} - \frac{5x-1}{14} = \frac{x-7}{10} - \frac{5x-6}{21}$

j) $25 - 7x = 41x - 119$

k) $3x + 100 = 5 \cdot (200 - 3x)$

l) $7 \cdot (x - 18) = 3 \cdot (x - 14)$

m) $\frac{x}{2} - \frac{3x}{4} + \frac{5x}{6} = 14$

n) $\frac{4x}{15} - \frac{2x-5}{20} = \frac{2x-1}{5}$

ñ) $\frac{3x-17}{8} - \frac{1-4x}{13} = \frac{1-x}{4} - \frac{9-x}{6}$

o) $2(2 - 3x) - 3(3 - 2x) = 4(x + 1) + 3(4 - 5x)$

4. Solve these equations:

a) $3x-2(1-2x)+1=3(x-2)+2x-1$

b) $3-\frac{1-2x}{3}+\frac{x}{9}=\frac{5-3x}{2}+x$

c) $x+\frac{x}{2}+\frac{x}{3}+\frac{x}{4}=1$

d) $\frac{2-x}{3}-3x=5-\frac{x-1}{2}$

e) $2x-[(3-x)-2(1+x)]=3x+5$

f) $\frac{x-1}{2}+\frac{x}{3}=\frac{x+1}{4}+\frac{3x-1}{6}$

g) $\frac{10x+9}{8}+\frac{x}{3}=3x+4$

h) $\frac{x-4}{5}-\frac{3(2x+1)}{2}=x$

i) $\frac{2}{3}(x-1)-\frac{1}{2}(3-x)=4x$

j) $\frac{2(5x-1)}{5}-\frac{3(4-x)}{2}=1$

Quadratic equations

5. Solve the following quadratic equations, without using the formula:

a) $3x^2-12x=0$ b) $x-3x^2=0$ c) $2x^2-5x=0$ d) $2x^2-8=0$

e) $9x^2-25=0$ f) $4x^2+100=0$ g) $16x^2=100$

6. Solve:

a) $x^2+4x-21=0$ b) $x^2+9x+20=0$ c) $9x^2-12x+4=0$

d) $x^2+x+3=0$ e) $4x^2+28x+49=0$ f) $x^2-2x+3=0$

g) $4x^2-20x+25=0$ h) $-2x^2+3x+2=0$

7. Solve the following quadratic equations:

a) $(2x+1)(x-3)=(x+1)(x-1)-8$

b) $(2x-3)(2x+3)-x(x+1)-5=0$

c) $(2x+1)^2=4+(x+2)(x-2)$

d) $(x+4)^2-(2x-1)^2=8x$

8. Solve the following quadratic equations:

a) $\frac{2x^2-1}{2}-\frac{x-1}{3}=\frac{1-x}{6}$

b) $\frac{x(x+3)}{2}-\frac{(x+1)^2}{3}+\frac{1}{3}=0$

c) $x(2x+1)-\frac{(x-1)^2}{2}=3$

9. Some of the following equations have 2 solutions, some have infinite solutions and some do not have solution. Find them.

a) $4(2x+1)-3(x+3)=5(x-2)$ b) $(x-3)+1=3(x-1)-(2+x)$

c) $\frac{3x+1}{2}=2x-\frac{x-1}{2}$ d) $x+\frac{2x-7}{4}=2x+\frac{1-x}{2}$

10. The sum of three consecutive natural numbers is equal to five times the lowest number minus 11 units. Which are the numbers?

11. Find out a number so that when adding it to its half, the result is the same as subtracting 6 to the 9/5 of that number.

12. Find out three odd numbers whose sum is 117.

13. Calculate the length of the sides of a rectangle whose perimeter is 82 cm and whose base is 8 cm longer than its height.

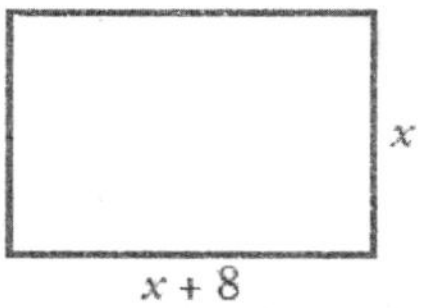

14. I have paid 14.30 € for a pen, a notebook and a book. If the price of the book is 5 times the price of the notebook, and the notebook costs twice as much as the pen, what is their prices?

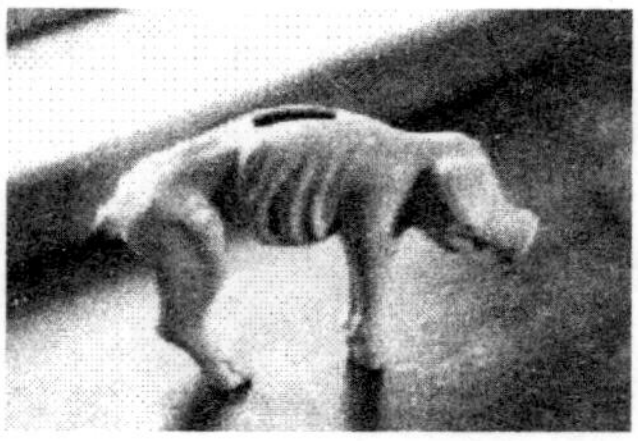

15. I had a quantity of money in the bank. I got 1/7 of it to buy something. After that, I put into my account 2/15 of the remaining, and, even now, I have 12 € less than the initial quantity. What was the initial quantity?

16. John is twice as old as his friend Peter. Peter is 5 years older than Alice. In 5 years, John will be three times as old as Alice. How old is Peter now?

17. The length and width of a rectangle are $(x+4)$ cm and x cm respectively. If the sum of its area plus a square with side x is 96 cm^2, find x.

18. The sum of the square of a number and six times the number is 91. Find the two possible values of the number.

19. The length of a rectangular lawn is 5 meters longer than its width. If the area of the lawn is $176m^2$, find the dimensions of the lawn.

20. Two positive numbers have a difference of 3 and their squares have a sum of 317. One of the numbers is x. Write down an equation with x based on this information and solve it to find the two numbers.

21. The legs of a right-angled triangle are (2x+1) and (2x+3)cm long. If the area of this triangle is 13 cm^2, calculate the value of x.

22. Arthur has a bag containing 28 candies, some of them are mints and the others are lemon. If the number of mints is three times the number of lemon candies, how many candies of each kind can Arthur eat?

23. John kept 4/5 of the oranges that were in the box. Mary took half of the oranges that were left in the box, leaving only 5 oranges. How many oranges did the box contain originally?

24. Rachel and Sarah have 45 biscuits between them. Sarah tells Rachel: "if you give me 5 biscuits I will have twice as many biscuits as you". How many biscuits does each of them have?

25. Two cyclists leave the same point, at the same time in the same direction. The first one pedals at a constant speed of 35 km/h and the second one at 25 km/h. How long will it take them to be 15 km apart?

26. Two cars leave the same point at the same time, but in opposite directions. The first one is driving at a constant speed of 85 km/h, and the second one at a constant speed of 90 km/h.

a) How many kilometers apart will they be in two hours?

b) How long will it take them to be 300 km apart?

27. Find the dimensions of a rectangular lawn knowing that the perimeter is 180 m long, and one side is 20 m longer than the other side.

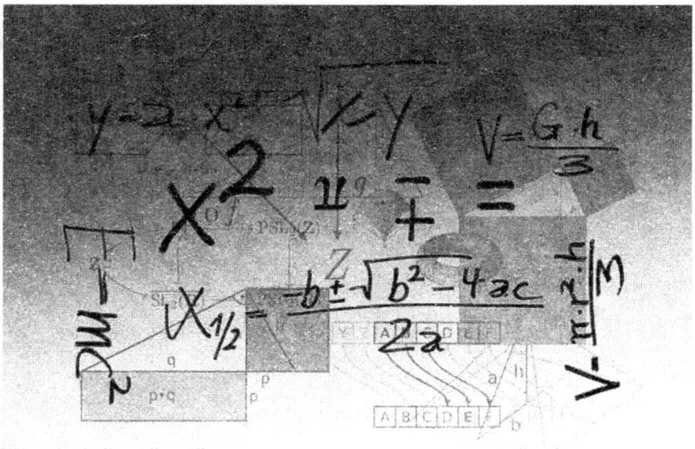

Unit 6.- Systems of equations

1. Linear equations with two unknowns

A ***linear equation with two unknowns*** is an equation like this:

$$ax + by = c$$

Numbers a and b are named ***coefficients*** of the unknowns x and y.

Number c is named ***independent term.***

One of these equations have infinite solutions, because you can introduce infinite values in one of the unknowns, and calculate the other one. In fact, you can build a chart with the solutions or draw a linear graph.

Example: For equation 2x + y = 30, complete the following chart and draw its linear graph.

x	1	2	3	5	10	12
y						

Solution:

- For $x = 1$, substituting in the equation: $2\cdot 1 + y = 30 \rightarrow y = 28$
- For x = 2, substituting in the equation: $2\cdot 2 + y = 30 \rightarrow y = 26$

ollowing chart shows some pairs of solutions for this equation:

x	1	2	3	5	10	12
y	28	26	24	20	10	6

Each of these pairs of values (x, y) is a ***solution of the equation.*** Equations with two onknown have infinite solutions and can be graphically represented by a line. If you represent above points in a graph, you will obtain the following:

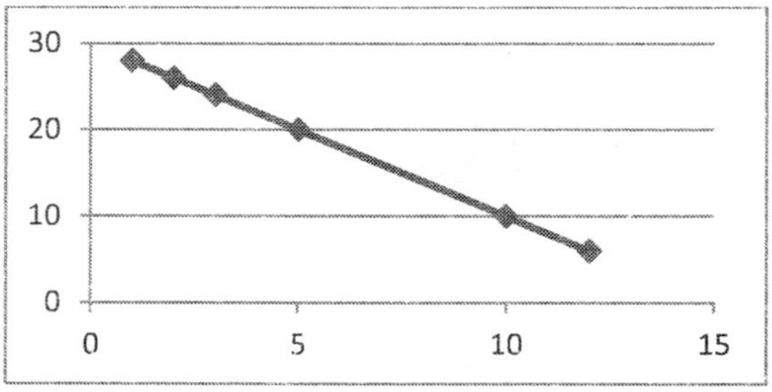

2. Solution of a system of equations

- A pair of numbers (x_0, y_0), are said to be ***solution*** of a equation with two unknowns

$$ax + by = c$$

if when substituting x by x_0 and y by y_0, equality is verified.

- A pair of numbers (x_0, y_0), are said to be ***solution*** of a ***system of linear equations*** with two unknowns

$$\begin{cases} ax + by = c \\ a'x + b'y = c' \end{cases}$$

if when substituting x by x_0 and y by y_0, both equalities are verified.

- Two or more systems of linear equations are said to be ***equivalent*** if they have the same solution.

Systems of equations are also named ***simultaneous equations***.

Example: Check which of the following pairs of numbers are solution of the equation:

$4x - 3y = 12$

a) $(x = 6, y = 4)$ b) $(x = 6, y = 12)$ c) $(x = 0, y = -4)$

Solution:

a) $4 \cdot 6 - 3 \cdot 4 = 24 - 12 = 12 \rightarrow$ $(x = 6, y = 4)$ is a solution of the equation.

b) $4 \cdot 6 - 3 \cdot 12 = 24 - 36 = -12 \rightarrow$ $(x = 6, y = 12)$ is not a solution of the equation.

c) $4 \cdot 0 - 3(-4) = 0 + 12 = 12 \rightarrow$ $(x = 0, y = -4)$ is a solution of the equation.

Example: Check if some of the following pairs $(x = -1, y = 4)$ and $(x = 7, y = 8)$ are solution of the following system of linear equations:

$$\begin{cases} -6x + 5y = 26 \\ x - 2y = -9 \end{cases}$$

Solution:

a) $(x = -1, y = 4)$ $\begin{cases} -6(-1) + 5 \cdot 4 = 6 + 20 = 26 \rightarrow OK!!! \\ (-1) - 2 \cdot 4 = -1 - 8 = -9 \rightarrow OK!!! \end{cases}$ So, $(x = -1, y = 4)$ is a solution.

b) $(x = 7, y = 8)$ $\begin{cases} -6 \cdot 8 + 5 \cdot 8 = -48 + 40 = -8 \neq 26 \rightarrow Not\ \ solution \\ 7 - 2 \cdot 8 = 7 - 16 = -9 \rightarrow OK!!! \end{cases}$

Exercises

1. Check if some of the following pairs are solution of the following system of equations:

$\begin{cases} -2x + 4y = 18 \\ 3x - 2y = 5 \end{cases}$ a) $(x = -1, y = 4)$ b) $(x = 7, y = 8)$

2. Check if some of the following pairs are solution of the following system of equations:

$\begin{cases} 5x + y = 43 \\ 3x + y = 1 \end{cases}$ a) $(x = -1, y = 4)$ b) $(x = 7, y = 8)$

3. Graphical solution of a system of equations

Example: Represent the linear plot of equations $2x - y = 6$, $x + y = 0$. Find in the graph the solution of both equations. Check the solution by substituting in both equations.

Solution:

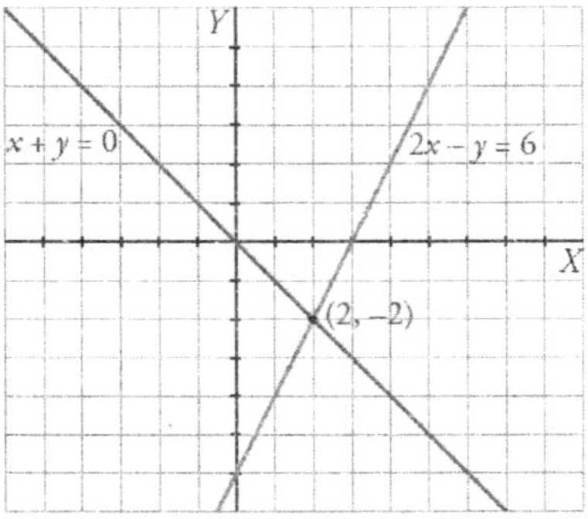

Common solution to both equations is $x = 2$, $y = -2$. Point (2, –2).

Check your solution!!! $\begin{cases} 2x - y = 6 \rightarrow 2 \cdot 2 - (-2) = 4 + 2 = 6 \rightarrow OK!!! \\ x + y = 0 \rightarrow 2 + (-2) = 2 - 2 = 0 \rightarrow OK!!! \end{cases}$

Exercises

3. Represent the following pairs of linear plots corresponding to each system of linear equations, find their solutions, and decide if they are equivalent.

a) $\begin{cases} x - 2y = -2 \\ 2x + 2y = 8 \end{cases}$ b) $\begin{cases} y - x = 0 \\ y = 2 \end{cases}$

4. Methods to solve a system of linear equations

There are three methods to solve systems of linear equations:

- Substitution method
- Equalization method
- Reduction method

4.1. Substitution method

We are going to describe ***substitution method*** with an example.

Example: Solve the following system by using the substitution method.

$$\begin{cases} x - 2y = -2 \\ 2x + 2y = 8 \end{cases}$$

<u>Solution:</u> You must follow the following steps:

Step 1: Isolate one unknown, where easier.	➡	In our case, easier unknown is "x" in the first equation: x = - 2 + 2y
Step 2: Substitute this expression in the <u>other</u> equation.	➡	2x + 2y = 8. As x = -2+2y, then 2·(-2 + 2y) + 2y = 8
Step 3: Solve this new equation with only one unknown.	➡	2·(-2 + 2y) + 2y = 8 - 4 + 4y + 2y = 8 4y + 2y = 4 + 8 6y = 12 $y = \frac{12}{6} = 2$ $\boxed{y = 2}$
Step 4: In order to obtain the value of the other unknown, we substitute known value in any of the two given equations. We choose it freely, but it is a good idea looking for the easiest one.	➡	x = - 2 + 2y As y = 2, then x = - 2 + 2·2 = - 2 + 4 = $\boxed{2}$ So, solution is $\boxed{(x = 2, y = 2)}$.
Step 5: Check your solution.	➡	$\begin{cases} x - 2y = -2 \rightarrow 2 - 2\cdot 2 = 2 - 4 = -2 \rightarrow OK!!! \\ 2x + 2y = 8 \rightarrow 2\cdot 2 + 2\cdot 2 = 4 + 4 = 8 \rightarrow OK!!! \end{cases}$

Exercises	
	4. Solve the following systems by using the *substitution* method. a) $\begin{cases} x+3y=5 \\ 5x+7y=13 \end{cases}$ b) $\begin{cases} 6x+3y=0 \\ 3x-y=3 \end{cases}$ c) $\begin{cases} 3x+9y=4 \\ 2x+3y=1 \end{cases}$ d) $\begin{cases} x-4y=11 \\ 5x+7y=1 \end{cases}$ **5.** Solve the following systems by using the *substitution* method. a) $\left.\begin{matrix} 3x+2y=23 \\ x+y=8 \end{matrix}\right\}$ b) $\left.\begin{matrix} 3x+2y=14 \\ 2x-5y=-16 \end{matrix}\right\}$ c) $\left.\begin{matrix} x+3y=4 \\ 2x-y=1 \end{matrix}\right\}$ d) $\left.\begin{matrix} 3x-y=17 \\ 2x+y=8 \end{matrix}\right\}$

4.2. Equalization method

We are going to describe ***equalization method*** with an example.

Example: Solve the following system by using the equalization method: $\begin{cases} 3x-2y=1 \\ 2x+5y=7 \end{cases}$

Solution: You must follow the following steps:

Step 1: Isolate one unknown (choose freely) in both equations. As you are free, choose the easiest one.	➡	We are going to isolate the "x": - First equation: $3x-2y=1 \rightarrow 3x=2y+1 \rightarrow \boxed{x=\frac{2y+1}{3}}$ - Second equation: $2x+5y=7 \rightarrow 2x=-5y+7 \rightarrow \boxed{x=\frac{-5y+7}{2}}$
Step 2: Build an equality with both expressions.	➡	$\frac{2y+1}{3}=\frac{-5y+7}{2}$
Step 3: Solve this new equation with only one unknown.	➡	LCM(3,2) = 6, we must mutiply all the terms by 6. $2(2y+1)=3(-5y+7)$ $4y+2=-15y+21$ $4y+15=-2+21$ $19y=19 \rightarrow y=\frac{19}{19}=1$ $\boxed{y=1}$

Step 4: In order to obtain the value of the other unknown, we substitute known value in any of the two given equations. We choose it freely, but it is a good idea looking for the easiest one.	➡	$2x + 5y = 7$ As $y = 1$, then $2x + 5\cdot 1 = 7$ $2x + 5 = 7$ $2x = 7 - 5 = 2$ $x = \frac{2}{2} = 1$ So, solution is $\boxed{(x = 1,\ y = 1)}$.
Step 5: Check your solution.	➡	$\begin{cases} 3x - 2y = 1 \to 3\cdot 1 - 2\cdot 1 = 3 - 2 = 1 \to OK!!! \\ 2x + 5y = 7 \to 2\cdot 1 + 5\cdot 1 = 2 + 5 = 7 \to OK!!! \end{cases}$

Exercises

6. Solve the following systems by using the *equalization* method.

a) $\begin{cases} x + 3y = 5 \\ 5x + 7y = 13 \end{cases}$ b) $\begin{cases} 6x + 3y = 0 \\ 3x - y = 3 \end{cases}$ c) $\begin{cases} 3x + 9y = 4 \\ 2x + 3y = 1 \end{cases}$ d) $\begin{cases} x - 4y = 11 \\ 5x + 7y = 1 \end{cases}$

7. Solve the following systems by using the *equalization* method.

a) $\left.\begin{matrix} 7x - 5y = 52 \\ 2x + 5y = 47 \end{matrix}\right\}$ b) $\left.\begin{matrix} 2x - y = 5 \\ x + 2y = 5 \end{matrix}\right\}$ c) $\left.\begin{matrix} 3x - y = 7 \\ 2x + 3y = 1 \end{matrix}\right\}$ d) $\left.\begin{matrix} 7x + 4y = 3 \\ 5x - 3y = 8 \end{matrix}\right\}$

4.3. Reduction method

Reduction method is based on the fact that there are operations that can be done on a system of equations without changing its solution. These operations are named ***equivalence operations***, as they transform a system into another different one, but equivalent to original one:

- Adding both equations.
- Multiplying one equation by a real number and, after, adding.
- Multiplying both equations by real numbers and, after, adding.

Equivalence operation

$\begin{cases} --Eq.\ 1-- \\ --Eq.\ 2-- \end{cases}$

→ Adding both equations. → $\begin{cases} Eq.\ 1\ or\ Eq.2 \\ Eq.\ 1+Eq.2 \end{cases}$

→ Multiplying one equation by a real number and, after, adding. → $\begin{cases} Eq.\ 1\ or\ Eq.2 \\ a\cdot Eq.\ 1+Eq.2 \end{cases}$

→ Multiplying both equations by real numbers and, after, adding. → $\begin{cases} Eq.\ 1\ or\ Eq.2 \\ a\cdot Eq.\ 1+b\cdot Eq.2 \end{cases}$

We are going to describe ***reduction method*** with examples.

Example: Solve the following systems by using the reduction method:

a) $\begin{cases} 3x-2y=1 \\ 5x+2y=7 \end{cases}$ b) $\begin{cases} 3x+y=1 \\ 5x+y=3 \end{cases}$ c) $\begin{cases} 3x+2y=1 \\ 5x+y=4 \end{cases}$ d) $\begin{cases} 3x+5y=1 \\ 2x+3y=1 \end{cases}$

Solution:

a) $\begin{cases} 3x-2y=1 \\ 5x+2y=7 \end{cases}$

Notice if you add both equations, terms containing unknown *"y"* will be cancellated.

$$\begin{cases} 3x-2y=1 \\ 5x+2y=7 \end{cases}$$

Addition: 8x = 8 Solving this equation is easy, being x = 1.

In order to obtain the value of the other unknown, we substitute known value in any of the two given equations. We choose it freely, but it is a good idea looking for the easiest one.

5x + 2y = 7 As x = 1, then

5·1 + 2y = 7

5 + 2y = 7

2y = 7 – 5 = 2

$x = \frac{2}{2} = 1$

So, solution is (x = 1, y = 1).

Finally, check your solution. $\begin{cases} 3x-2y=1 \to 3\cdot 1-2\cdot 1=3-2=1 \to OK!!! \\ 5x+2y=7 \to 5\cdot 1+2\cdot 1=5+2=7 \to OK!!! \end{cases}$

b) $\begin{cases} 3x + y = 1 \\ 5x + y = 3 \end{cases}$

Notice if you add both equations, none of the unknowns are being cancelated. We need a (-) sign before any of the *"y"* unknowns. So, we must multiply one of the equations by (-1).

$$\begin{cases} 3x + y = 1 \ \underrightarrow{\textit{Multiply by } (-1)} \\ 5x + y = 3 \ \underrightarrow{\textit{Leave}} \end{cases} \qquad \begin{cases} -3x - y = -1 \\ 5x + y = 3 \end{cases}$$

Now, if you add both equations, terms containing unknown *"y"* will be canceled.

$$\begin{cases} -3x - y = -1 \\ 5x + y = 3 \end{cases}$$

Addition: 2x = 2 Solving this equation, x = $\boxed{1}$.

In order to obtain the value of the other unknown, we substitute known value in any of the two given equations. We choose it freely, but it is a good idea looking for the easiest one.

5x + y = 3 As x = 1, then

5·1 + y = 3

5 + y = 3

y = 3 – 5 = $\boxed{(-2)}$

So, solution is $\boxed{(x = 1,\ y = (-2))}$.

Finally, check your solution. $\begin{cases} 3x + y = 1 \rightarrow 3\cdot 1 + (-2) = 3 - 2 = 1 \rightarrow OK!!! \\ 5x + y = 3 \rightarrow 5\cdot 1 + (-2) = 5 - 2 = 3 \rightarrow OK!!! \end{cases}$

c) $\begin{cases} 3x + 2y = 1 \\ 5x + y = 4 \end{cases}$

Notice if you add both equations, none of the unknowns are being canceled. We need a (-2) before

"y" unknown in second equation. So, we must multiply second equation by (-2).

$$\begin{cases} 3x + 2y = 1 \ \underrightarrow{\textit{Leave}} \\ 5x + y = 4 \ \underrightarrow{\textit{Multiply by } (-2)} \end{cases} \qquad \begin{cases} 3x + 2y = 1 \\ -10x - 2y = -8 \end{cases}$$

Now, if you add both equations, terms containing unknown *"y"* will be canceled.

$$\begin{cases} 3x + 2y = 1 \\ -10x - 2y = -8 \end{cases}$$

Addition: -7x = -7 Solving this equation, x = $\boxed{1}$.

In order to obtain the value of the other unknown, we substitute known value in any of the two given equations. We choose it freely, but it is a good idea looking for the easiest one.

5x + y = 4 As x = 1, then

5·1 + y = 4 → 5 + y = 4 → y = 4 – 5 = $\boxed{(-1)}$ So, solution is $\boxed{(x = 1,\ y = (-1))}$.

d) $\begin{cases} 3x + 5y = 1 \\ 2x + 3y = 1 \end{cases}$

Notice if you add both equations, none of the unknowns are being cancelled. Moreover, even multiplying any of the equations by a real number, we would not be able to cancel any unknown.

When this happens, you must choose one of the unknowns and *cross-multiplying*.

For example, if we work on "x", we can multiply by 2 above and by (-3) below:

$$\begin{cases} 3x + 5y = 1 \ \underrightarrow{\textit{Multiply by } 2} \\ 2x + 3y = 1 \ \underrightarrow{\textit{Multiply by } (-3)} \end{cases} \quad \begin{cases} 6x + 10y = 2 \\ -6x - 9y = -3 \end{cases}$$

Now, if you add both equations, terms containing unknown *"x"* will be cancelled.

$$\begin{cases} 6x + 10y = 2 \\ -6x - 9y = -3 \end{cases}$$

Addition: y = - 1 So, y = $\boxed{(-1)}$.

In order to obtain the value of the other unknown, we substitute known value in any of the two given equations. We choose it freely, but it is a good idea looking for the easiest one.

2x + 3y = 1 As x = 1, then

2x + 3 · (-1) = 1

2x – 3 = 1

2x = 1 + 3 = 4 → 2x = 4 → x = $\boxed{2}$

So, solution is $(x=2,\ y=(-1))$.

Finally, check your solution. $\begin{cases} 3x+5y=1 \rightarrow 3\cdot 2+5\cdot(-1)=6-5=1 \rightarrow OK!!! \\ 2x+3y=1 \rightarrow 2\cdot 2+3\cdot(-1)=4-3=1 \rightarrow OK!!! \end{cases}$

Exercises

8. Solve the following systems by using the *reduction* method.

a) $\begin{cases} 3x+5y=11 \\ 4x-5y=38 \end{cases}$ b) $\begin{cases} x+3y=5 \\ 5x+7y=13 \end{cases}$ c) $\begin{cases} x-4y=11 \\ 5x+7y=1 \end{cases}$ d) $\begin{cases} 6x+3y=0 \\ 3x-y=3 \end{cases}$

9. Solve the following systems by using the *reduction* method.

a) $\left.\begin{matrix} 7x-5y=52 \\ 2x+5y=47 \end{matrix}\right\}$ b) $\left.\begin{matrix} 2x-y=5 \\ x+2y=5 \end{matrix}\right\}$ c) $\left.\begin{matrix} 3x-y=7 \\ 2x+3y=1 \end{matrix}\right\}$ d) $\left.\begin{matrix} 7x+4y=3 \\ 5x-3y=8 \end{matrix}\right\}$

10. Solve the following systems by any method:

a) $\left.\begin{matrix} x+y=2 \\ x-y=6 \end{matrix}\right\}$ b) $\left.\begin{matrix} x+y=12 \\ x-y=2 \end{matrix}\right\}$ c) $\left.\begin{matrix} x+y=5 \\ -x+y=1 \end{matrix}\right\}$ d) $\left.\begin{matrix} 3x-4y=-6 \\ x+2y=8 \end{matrix}\right\}$

e) $\left.\begin{matrix} 3x-y=17 \\ 2x+y=8 \end{matrix}\right\}$ d) $\left.\begin{matrix} x+2y=7 \\ 7x-4y=13 \end{matrix}\right\}$

5. Problems with systems of equations

Although lots of problems in algebra can be solved by using an equation with an only unknown, they become much easier if solved with a system of equations with two unknowns. We are seeing some examples.

Example: If 4 apples and 2 oranges cost 1 € and 2 apples and 3 oranges equals 0.70 €, how much does each apple and each orange cost?

Solution:

Setting	**System of equation**	**Solution**
x = price of one apple. y = price of one orange.	$\begin{cases} 4x+2y=1 \\ 2x+3y=0.70 \end{cases}$	Feel free to choose one of the methods. Remember to check your solution. - x = one apple costs 0.2 € - y = one orange costs 0.1 €.

Example: Find the value of two numbers if their sum is 12 and their difference is 4.

Solution:

Setting	System of equation	Solution
x = highest number.	$\begin{cases} x+y=12 \\ x-y=4 \end{cases}$	Feel free to choose one of the methods.
y = lowest number.		Remember to check your solution.
		Our numbers are 8 and 4.

Exercises

11. The sum of the digits of a certain two-digit number is 7. Reversing its digits, increases the number by 9. What is the number?

12. The school that Stefan goes to is selling tickets to a choral performance. On the first day of tickets sales the school sold 3 senior tickets and 1 child ticket for a total of $38. The school took in $52 on the second day by selling 3 senior tickets and 2 child tickets. Find the price of a senior ticket And a child ticket.

6. Number of solutions of a system of linear equations

If you remember, linear equations, studied in the previous unit, a linear equation can be classified as:

- ***Determined compatible:*** it has an only solution.
- ***Undetermined compatible:*** it has infinite solutions.
- ***Non-compatible:*** it has not solution.

The same classification is valid for systems of linear equations. Let´s see it with examples.

Example: Represent the linear plots corresponding to each system of equations and find their solution.

a) $\begin{cases} 2x + y = 7 \\ 4x + 2y = 14 \end{cases}$ b) $\begin{cases} 2x + y = 7 \\ 2x + y = 0 \end{cases}$ c) $\begin{cases} 2x + y = 7 \\ -2x + 5y = 10 \end{cases}$

Solution: Feel free to give values to "x", calculate corresponding "y-values", and you must obtain plots like following:

a) $\begin{cases} 2x + y = 7 \\ 4x + 2y = 14 \end{cases}$

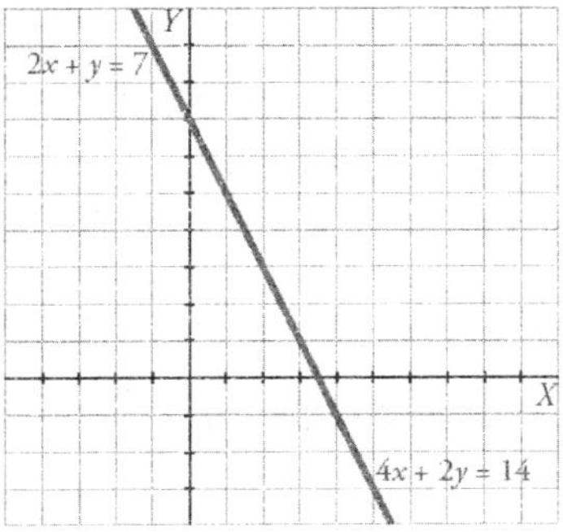

As you can see, both linear plots are *coincident*, so, they have infinite common points.

This system has ***infinite solutions***.

This system is ***indetermined compatible***.

b) $\begin{cases} 2x + y = 7 \\ 2x + y = 0 \end{cases}$

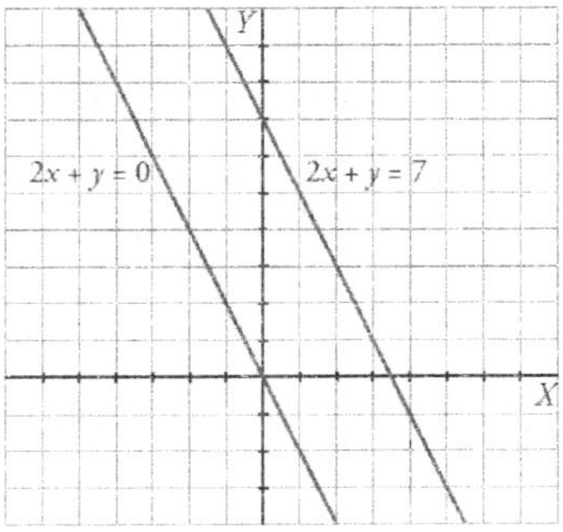

As you can see, both linear plots are *paralel*, so, they do not have any common point.

This system has ***not solution***.

This system is ***non-compatible***.

c) $\begin{cases} 2x + y = 7 \\ -2x + 5y = 10 \end{cases}$

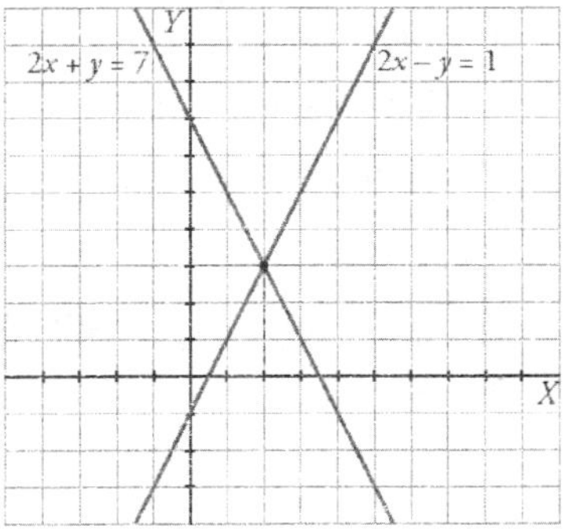

As you can see, both linear plots are *secant*, so, they have one common point.

This system has ***one solution***.

This system is ***determined compatible***.

QUICK IDENTIFICATION

Pay attention to the following:

a) $\begin{cases} 2x + y = 7 \\ 4x + 2y = 14 \end{cases} \rightarrow \frac{2}{4} = \frac{1}{2} = \frac{7}{14}$ b) $\begin{cases} 2x + y = 7 \\ 2x + y = 0 \end{cases} \rightarrow \frac{2}{2} = \frac{1}{1} \neq \frac{7}{0}$ c) $\begin{cases} 2x + y = 7 \\ -2x + 5y = 10 \end{cases} \rightarrow \frac{2}{-2} \neq \frac{1}{5}$

As a general rule: For a general system oflinear equations $\begin{cases} Ax + By = C \\ A'x + B'y = C' \end{cases}$

$\frac{A}{A'} = \frac{B}{B'} = \frac{C}{C'}$	$\frac{A}{A'} = \frac{B}{B'} \neq \frac{C}{C'}$	$\frac{A}{A'} \neq \frac{B}{B'}$
Indetermined compatible	Non-compatible	Determined compatible

Exercises

13. Only by looking at them, indicate if following systems have an only solution, infinite solutions or they do not have any solution.

a) $\begin{cases} 2x - y = 1 \\ 4x - 2y = 8 \end{cases}$ b) $\begin{cases} x - 2y = 5 \\ 2x - 4y = 10 \end{cases}$ c) $\begin{cases} 5x + 2y = -1 \\ 4x - y = 7 \end{cases}$ d) $\begin{cases} x - 2y = 5 \\ 2x - 4y = -3 \end{cases}$

14. Complete the following systems so that solution for the first one is ($x = 3$, $y = -2$), second one is non-compatible, and third and fourth are undetermined compatible.

a) $\begin{cases} 3x + 2y = \dots \\ \dots - y = 8 \end{cases}$ b) $\begin{cases} x + y = 5 \\ 2x + 2y = \dots \end{cases}$ c) $\begin{cases} 3x - 2y = 4 \\ 6x - 4y = \dots \end{cases}$ d) $\begin{cases} -x + 2y = 7 \\ \dots - 4y = \dots \end{cases}$

Review exercises

1. Check if x = 2 and y = (-1) is a solution for the following systems of equations:

a) $\begin{cases} 2x - y = -4 \\ 5x + y = -10 \end{cases}$ b) $\begin{cases} 3x - 4y = 10 \\ 4x + 3y = 5 \end{cases}$

2. a) Represent together the linear plots whose equations are:

2x + y = 3 x – y = 3

b) According to previous plot, find the solution of the system $\begin{cases} 2x + y = 3 \\ x - y = 3 \end{cases}$

3. Represent the plots and find the solution of the following systems of equations:

a) $\begin{cases} 3x - y = 1 \\ x + 2y = 5 \end{cases}$ b) $\begin{cases} 3x - y = 0 \\ 3x + y = -6 \end{cases}$ c) $\begin{cases} x + 3y = -5 \\ 2x - y = 4 \end{cases}$ d) $\begin{cases} 2x - 3y = -4 \\ x + 8y = -2 \end{cases}$

4. Solve by using the substitution method:

a) $\begin{cases} x + 3y = 0 \\ 2x + y = -5 \end{cases}$ b) $\begin{cases} 8x - 3y = -25 \\ x - 5y = -17 \end{cases}$ c) $\begin{cases} 7x - y = -6 \\ 4x + 3y = 3 \end{cases}$ d) $\begin{cases} 2x + 16 = 2y \\ 2y - 3x = 16 \end{cases}$

5. Solve by using the equalization method:

a) $\begin{cases} x = 4 \\ x - y = 6 \end{cases}$ b) $\begin{cases} x + 3y = -4 \\ x - 2y = 6 \end{cases}$ c) $\begin{cases} y = 6x \\ 7x = 2y - 5 \end{cases}$ d) $\begin{cases} 3x - 4y = -4 \\ 2x + y = -1 \end{cases}$

6. Solve by using the reduction method:

a) $\begin{cases} x + y = 0 \\ x - y = 2 \end{cases}$ b) $\begin{cases} 3x - y = 0 \\ 3x + y = -6 \end{cases}$ c) $\begin{cases} 4x - 3y = 2 \\ 2x + y = -4 \end{cases}$ d) $\begin{cases} x + 2y = 1 \\ 3x - y = 7 \end{cases}$

e) $\begin{cases} x - 3y = 1 \\ 3x + 6y = 2 \end{cases}$ f) $\begin{cases} 3x + 2y = 3 \\ x + y = 7/6 \end{cases}$

7. Solve the following systems by using the method you consider more suitable:

a) $\begin{cases} x - y = 1 \\ 4x - 3y = 8 \end{cases}$ b) $\begin{cases} 3x = 1 + y \\ 3 + 2y = 10x \end{cases}$ c) $\begin{cases} 2x + 5y = -1 \\ 4x - 3y = -2 \end{cases}$ d) $\begin{cases} 3x - 2y = 2 \\ x + 4y = -5/3 \end{cases}$

8. Solve the following systems by using the method you consider more suitable:

a) $\begin{cases} 2x + y = 0 \\ 5x - 3 = 9y - 3 \end{cases}$ b) $\begin{cases} \dfrac{x}{3} - \dfrac{y}{2} = 4 \\ \dfrac{x}{2} + \dfrac{y}{4} = 2 \end{cases}$ c) $\begin{cases} 2(3x - 2) = y - 1 \\ 3(x + y) + 2(x - y) = 8 \end{cases}$ d) $\begin{cases} x + \dfrac{y - 2}{4} = 1 \\ x - \dfrac{3}{2}y = 5 \end{cases}$

9. Observe the following systems and indcate if they have one solution, infinite sutions or they do not have solutions. Represent their corresponding linear plots and check your answers.

a) $\begin{cases} 2x - y = 1 \\ 4x - 2y = 8 \end{cases}$ b) $\begin{cases} x - 2y = 5 \\ 2x - 4y = 10 \end{cases}$ c) $\begin{cases} 5x + 2y = -1 \\ 4x - y = 7 \end{cases}$ d) $\begin{cases} x - 2y = 5 \\ 2x - 4y = -3 \end{cases}$

10. Complete the following systems of equations so that solution for the first one is ($x = 6$, $y = -1$), second one is non-compatible, and third and fourth are undetermined compatible.

a) $\begin{cases} x - 4y = \dots\dots \\ 2x\dots\dots = 13 \end{cases}$ b) $\begin{cases} 2x + y = 8 \\ 4x + 2y = \dots\dots \end{cases}$ c) $\begin{cases} 2x + y = 8 \\ 4x\dots\dots = \dots\dots \end{cases}$ d) $\begin{cases} 5x + 11y = \dots\dots \\ \dots\dots + 33y = 9 \end{cases}$

11. Represent the following systems and indicate if they are equivalent:

a) $\begin{cases} 2x + y = 3 \\ x - y = 3 \end{cases}$ b) $\begin{cases} y + 1 = 0 \\ 3x - 4y = 10 \end{cases}$

12. Solve the following systems by reduction:

a) $\begin{cases} 13x-8y=15 \\ 7x-14y=9 \end{cases}$ b) $\begin{cases} 9x-13y=54 \\ 11x-7y=22 \end{cases}$

13. Solve the following systems. Previously, simplify the equations:

a) $\begin{cases} \frac{2}{3}x+y=-1 \\ \frac{x+1}{3}+\frac{y-1}{6}=-1 \end{cases}$ b) $\begin{cases} \frac{x+1}{3}+y=1 \\ \frac{x-3}{4}+2y=1 \end{cases}$ c) $\begin{cases} \frac{2-x}{3}+\frac{3+y}{6}=2 \\ \frac{8-3x}{6}-\frac{2+y}{9}=2 \end{cases}$

d) $\begin{cases} \frac{x-1}{2}+\frac{y+1}{4}=1 \\ \frac{2x-1}{2}-\frac{2y+1}{6}=1 \end{cases}$

14. Solve the following systems, indicating if some of them are undetermined compatible or non-compatible:

a) $\begin{cases} 2x-5y=-2 \\ 3.25x-2.5y=8 \end{cases}$ b) $\begin{cases} 0.2x-1.7y=6.1 \\ 3x+2y=9 \end{cases}$ c) $\begin{cases} 3(x-1)+y=0 \\ 3(x+1)+y=-5 \end{cases}$

d) $\begin{cases} x+y=4-y \\ 3x-5=7-6y \end{cases}$

15. Find out two numbers whose sum is 160 and whose difference is 34.

16. Two pens and three notebooks cost 7.80 €; five pens and four notebooks cost 13.2 €. What are the prices of each product?

17. In a book shop, they have sold 45 books, some of them cost 32 € each, and some others, 28 € each. They got 1368 €. How many books of each kind were sold?

18. In a farm, there are rabbits and chickens, which equal to a total of 29 heads and 92 legs. How many rabbits and chickens are there?

19. An exam has 50 questions and I must answer all of them. Each correct answer adds 1 point for me and each wrong answer, will subtract 0.5 points. If my mark has been 24.5 points, how many correct and wrong answers have I had?

20. An olive oil company has bottled 2000 liters of oil into 1.5 L and 2 L bottles. If they have used 1100 bottles, how many of them are 1.5 L and 2 L?

21. Find two natural numbers whose sum is 154 and whose quotient is 8/3.

22. Find out two natural numbers whose sum is 140 and so that when dividing the highest one by the lowest one, the quotient is 2 and the remainder is 14.

23. The sum of the ages of a woman and her child is 56. Ten years ago, the mother´s age was five times the age of her child. What era their ages today?

24. Three years ago, Nuria´s age was twice her sister Marta. In 7 years, Nuria´s age will be the 4/3 of Marta´s one. Calculate their ages today.

25. The lowest base of an isosceles trapezium is 6 m long, and highest base is as long as two identical sides together. If its perimeter is 38 m, what is the length of each side?

26. If Álvaro gives Rita 4 of his CDs, she will have twice as many CDs as him. If Rita gives Álvaro 6 of her CDs, then he will have twice as many CDs as her. How many CDs do each one of them have?

27. The sum of the two digits of a number is 7. If we invert the order of the digits, we will obtain another number that is twice the original one plus 2 units. What was the original number?

28. Create a system of linear equations whose only solution is x = 2, y = (-1).

29. Look at the plots of lines r_1, r_2 and r_3. Answer without solving:

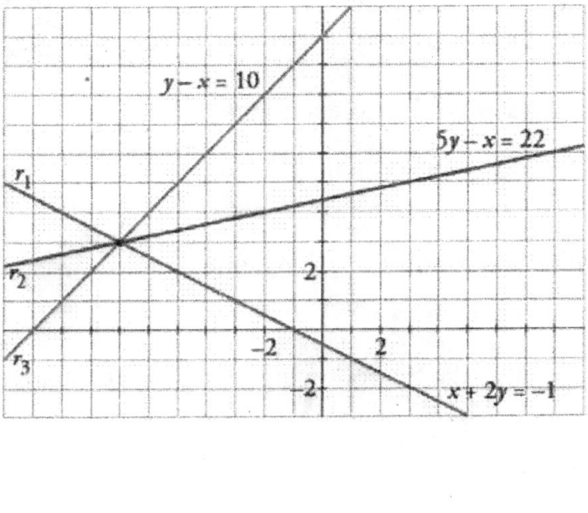

a) What is the solution of the systems: I) $\begin{cases} y-x=10 \\ x+2y=-1 \end{cases}$ II) $\begin{cases} 5y-x=22 \\ y-x=10 \end{cases}$

b) What is the solution of the system:

$$\begin{cases} y-x=10 \\ x+2y=-1 \\ 5y-x=22 \end{cases}$$

30. Look at the plots of lines r_1, r_2 and r_3. Answer without solving:

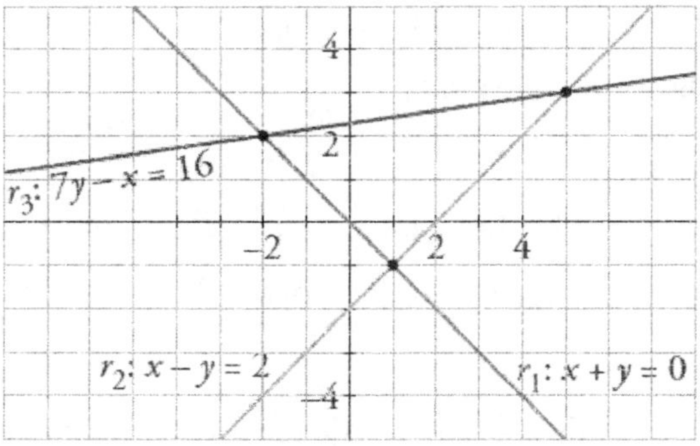

a) What is the solution of the systems: I) $\begin{cases} x+y=0 \\ x-y=2 \end{cases}$ II) $\begin{cases} x+y=0 \\ 7y-x=16 \end{cases}$ III) $\begin{cases} x-y=2 \\ 7y-x=16 \end{cases}$

b) What is the solution of the system:

$$\begin{cases} x+y=0 \\ x-y=2 \\ 7y-x=16 \end{cases}$$

31. Consider the following system of linear equations

$$\begin{cases} 3x+2y=5 \\ ax+by=15 \end{cases}$$

What must be the values of a and b that make it to have infinite solutions?

32. Consider the following system of linear equations

$$\begin{cases} 3x+2y=c \\ 6x+4y=d \end{cases}$$

Find the relationship between c and d that make this system non-compatible.

33. What is the value of m that make the following systems to be equivalent?

$$\begin{cases} 2x-3y=1 \\ x+y=8 \end{cases} \qquad \begin{cases} x-y=m \\ y=3 \end{cases}$$

ADVANCED

34. Solve by substitution:

$$\begin{cases} 2x - y = 2 \\ x^2 + y^2 = 52 \end{cases}$$

35. Solve by substitution:

a) $\begin{cases} 2x + y = 4 \\ x^2 - y = 7 \end{cases}$ b) $\begin{cases} y - x = 0 \\ 2x^2 + y^2 = 147 \end{cases}$ c) $\begin{cases} x - y = 2 \\ x^2 - y^2 = 16 \end{cases}$ d) $\begin{cases} x + y = 1 \\ 2x^2 - y^2 = 2 \end{cases}$

36. Difference of two natural numbers is 2, and difference of their squared is 20. Find them out.

37. Find out two natural numbers whose sum is 12 and whose sum of their squared is 80.

38. Diagonal of a rectangle is 15 cm and its perimeter is 42 cm. Calculate its sides.

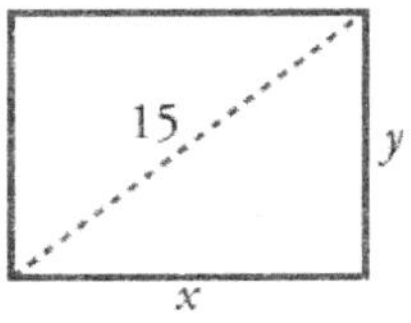

39. Perimeter of a rectangle is 68 cm and its area is 240 cm^2. Calculate its sides.

40. Perimeter of a isosceles triangle is 36 m. Height from the unlike side is 12m. Calculate the lengths of the like sides.

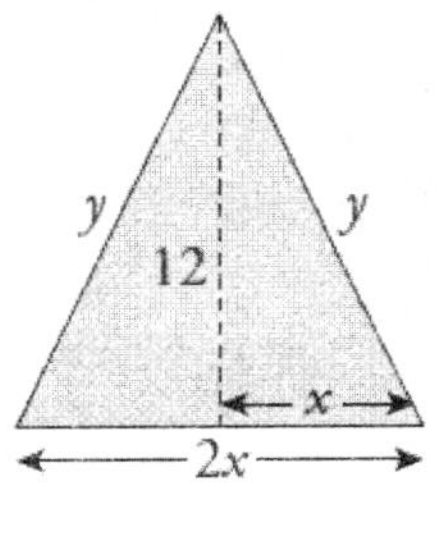

41. Sides of a triangle are 5 cm, 7 cm and 10 cm. Calculate the height on the longest side and the area of the triangle. *<u>Clue:</u> height h divides the triangle into two right triangles.*

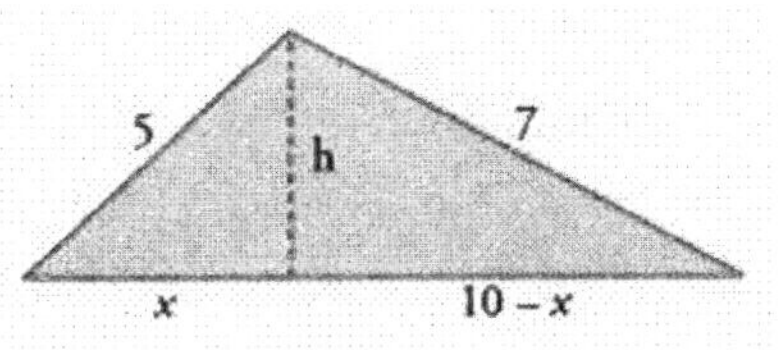

42. In a rectangle shaped field with perimeter 60 m, it is built a rectangle shaped garden, surrounded by a 2 m wide path. Calculate dimensions of the field, by knowing the area of the garden, 112 m^2.

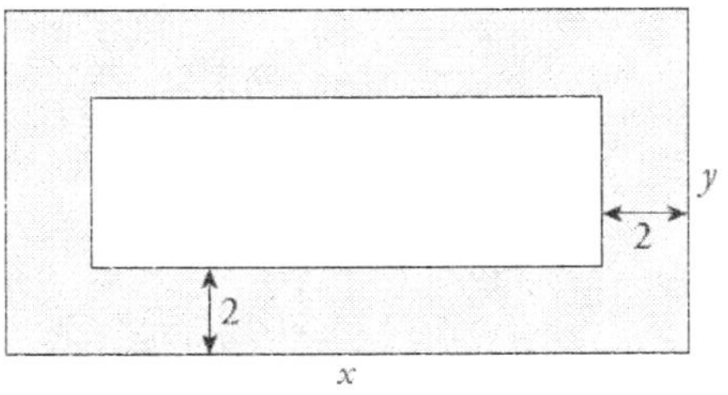

43. Some friends are sharing a 800 € prize, having everyone the same portion. Two of them decide not to take their portion and, in this way, the others receive 20 € more each. How many friends are they?

44. If base of a rectangle is decreased in 2 cm and height is increased in 4 cm, then it becomes a square. If base decreases in 4 cm and height increases in 4 cm, then its area decreases 12 cm^2. Calculate the sides of the rectangle.

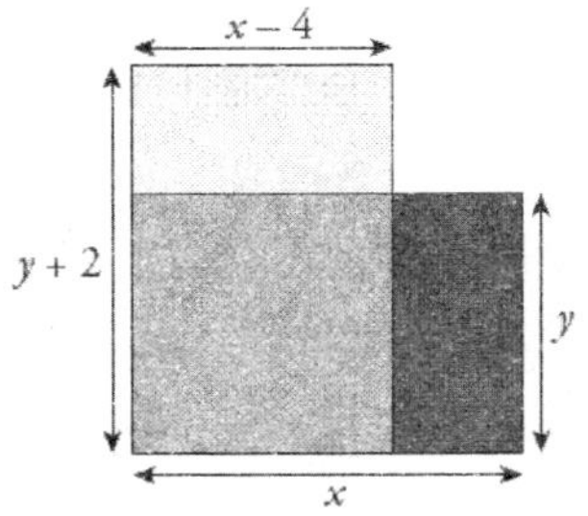

45. Solve this system of three equations and two unknowns and check graphically its solution.

$$\begin{cases} 2x - y = 1 \\ x - y = 5 \\ x + y = -1 \end{cases}$$

46. Solve these systems of three equations and three unknowns. Try expand your idea of substitution method.

a) $\begin{cases} x - 3 = 0 \\ 2x - 3y = 9 \\ x + y - z = 1 \end{cases}$ b) $\begin{cases} 3x - 2y + z = 4 \\ x + y = 5 \\ z - 3 = 1 \end{cases}$ c) $\begin{cases} x + y = z \\ x - y = z \\ x + z = -4 \end{cases}$ d) $\begin{cases} x + y = 5 \\ x + y + z = 3 \\ y + z = 2 \end{cases}$

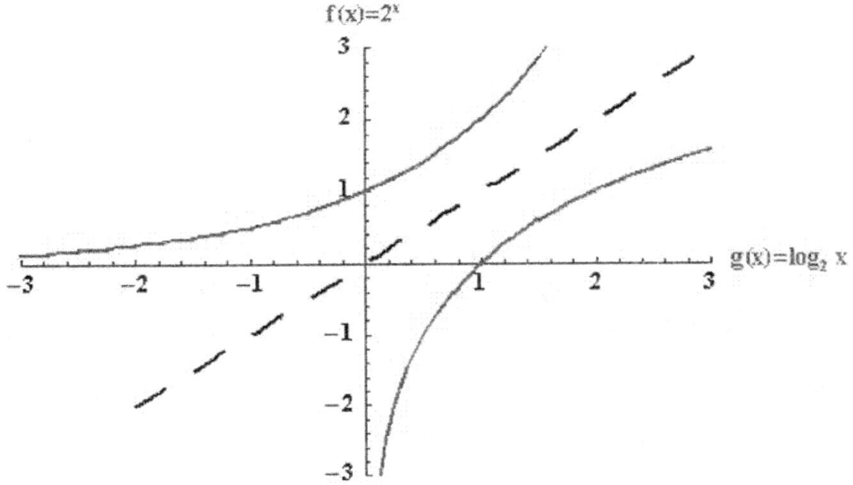

Unit 7.- Functions and graphs

1. Starting. Coordinates in a plane plot

Before, starting this unit about functions and graphs, we need to remember some concepts about them.

Example: Represent the following values in the real line: 2, 0, -1, 3, -2, 1, -3.

Consider two lines, one of them horizontal and the other one, vertical. They will be the reference axes to locate points in the plane $\mathbf{R}^2$. They are named ***X-axis*** or ***abscissa axis***, and ***Y-axis*** or ***ordinate axis***.

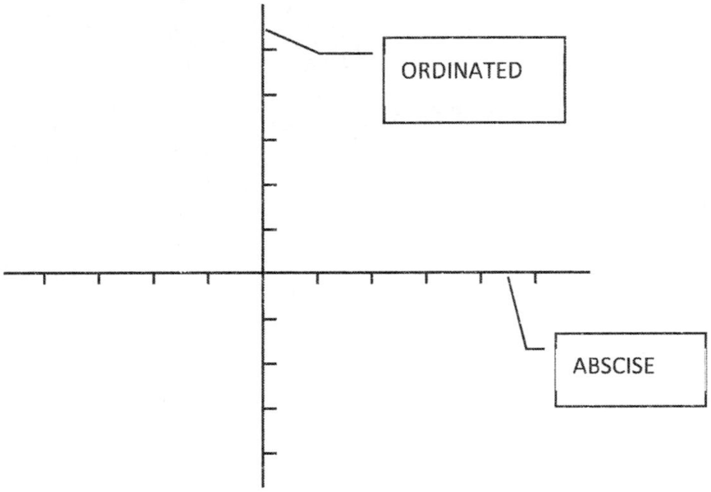

Location of points in the real plane, $\mathbf{R}^2$, is indicated by two numbers or ***coordinates***, ***(x, y).***

Example: Copy the coordinated axes and represent the following points:

A (2,3), B (4 ,1), C (-3,2), D (-4,-2), E (5,-1), F (3,-3), G (8, -6), H (-7, 5).

Example: A cyclist advances a distance of 70 km in 3 hours and he has been writing down times and distances in a table.

Distance (km)	0	10	20	30	40	50	60	70
Time (min)	0	30	60	90	120	140	160	180

Write down times in the abscise axis and distances in the ordinated axis. You do not need to use the same scale in both axes.

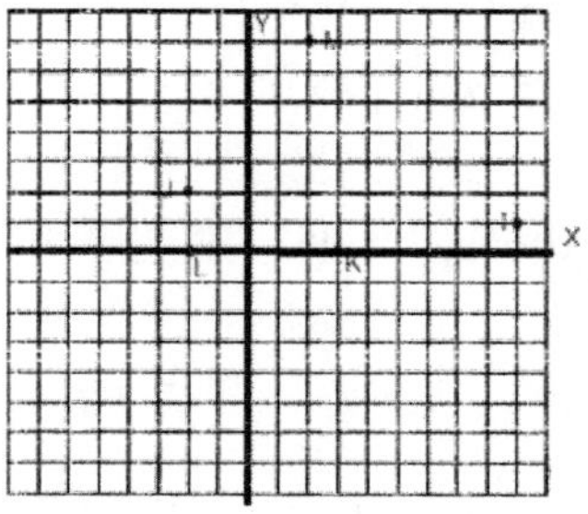

Example: Locate points A to H, and write down coordinates of points I to M.

A(3, 2) B(4, -5) C(-1, 3) D(-7, -2),

E(-5, 0) F(4, 0) G(0,-1) H(0, 1)

Example: About following graph, write down the images of -4, -3, -2, -1, 0, 1, 2, 3 and 4:

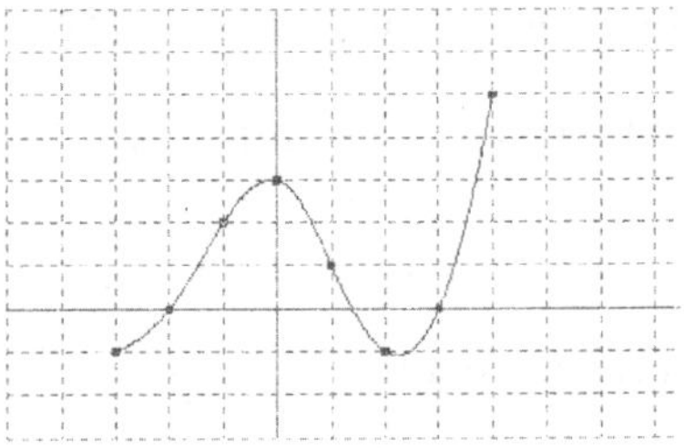

f(-4) = f(1) =

f(-3) = f(2) =

f(-2) = f(3) =

f(-1) = f(4) =

f(0) =

2. Dependence between magnitudes

a) Relationships given by tables

In a laboratory class, a student has measured temperature of a liquid while it was being heated. He wrote results in the following table:

x = time (min)	0	1	2	3	...
y = temperature (ºC)	20	24	28	32	...

Temperature of the liquid (*dependent variable*) depends on the time (*independent variable*).

b) Relationships given by a graph

Following graph represents variation of temperature of an ill person in a hospital during a whole day. From it, it is easy to determine the relationship between the time and the temperature of that person. For example, his lowest temperature has been reached at 2 p.m. (14:00 h) and highest one at 8 p.m. (20:00 h).

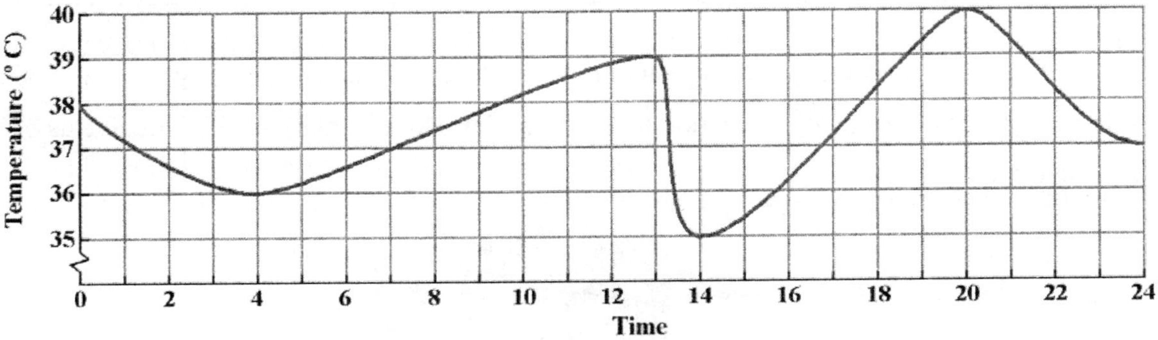

Graph gives us an intuitive view about the relationship between two magnitudes and the behavior of one magnitude, depending on the other one. Again, temperature (*dependent variable*) depends on the time in the day (*independent variable*).

c) Relationships given by formulas

Many of the relationships between magnitudes in Nature are given by algebraic expressions.

- Volume of a sphere depends on its side. The formula that relates both magnitudes is $V = \frac{4}{3}\pi a^3$.

- If price of 1 kg of oranges is 0.35 €, the cost *(C)* of a quantity of oranges depends on the number *(k)* of kilograms. The formula that relates both magnitudes is $C = 0.35k$.

Dependent variable is the one that depend on the other one, named, ***independent variable.*** Dependent variable is usually denoted with the letter ***y***, and independent variable is usually denoted with the letter ***x***.

3. Concepts concerning to functions

Above relationships have one common characteristic: each value of the independent variable corresponds an ***only*** value of the dependent variable. A relationship verifying this characteristic is named *function*.

A ***function*** is a relationship or correspondence between two magnitudes, so that each value of the independent variable *x*, corresponds ***an only value*** of the dependent variable *y*.

In order to indicate that a magnitude *(y)* depends or is a function of another one *(x)*, it is used the notation ***y = f(x)***, which is read *"y is a function of x"*.

Functions are like machines in which you introduce an element, *x*, and it gives you back another value, $y = f(x)$. For example, in function $f(x) = x^2$, we introduce *x* values, and it will give us back their squares.

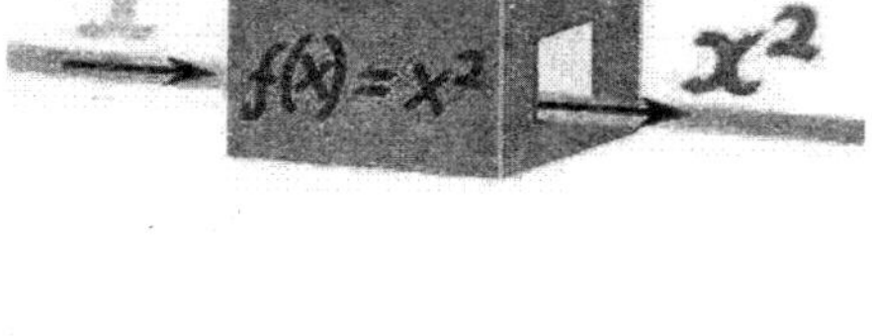

Calculator keys define functions by using formulas. For example, square root key, $\boxed{\sqrt{\ }}$ defines the function $y = \sqrt{x}$, or $f(x) = \sqrt{x}$.

For example, if we introduce in a calculator 25 and push $\boxed{\sqrt{\ }}$ we will see in the screen: 5. → This means 25 is a valid input and 5 is a valid output.

5 is said to be the *image* or *transformed* of 25 in the function $f(x)=\sqrt{x}$.
It is written *f(25) = 5.*

For example, if we introduce - 4 and push $\boxed{\sqrt{\ }}$ we will see in the screen: ERROR → This means - 4 is not a valid input for this function.
$y=\sqrt{-4}$ y is not a valid output for this function, there is not an input that
gives us back in the screen $\sqrt{-4}$

Example: Given the function *f* that joins each real number to its double:

a) Write its algebraic expression.

b) Calculate the images of 1, 2 and 3, denoted by *f(1)*, *f(2)* and *f(3)*.

Solution:

a) Algebraic expression of this function is denoted by $y = 2x$, or also by $f(x) = 2x$.

b) $f(1) = 2 \times 1 = \boxed{2}$; $f(2) = 2 \times 2 = \boxed{4}$; $f(3) = 2 \times 3 = \boxed{6}$.

1. Explain if the following relationships are functions, indicating independent (x) and dependent (y) variables:

a) Each real number is transformed in its cubic root.

b) Each integer number is transformed in its prime factors.

2. Explain if following relationships are functions, indicating independent (x) and dependent (y) variables: each real number is transformed in its third plus 1 unit.

3. Which of the following graphs represent a function?

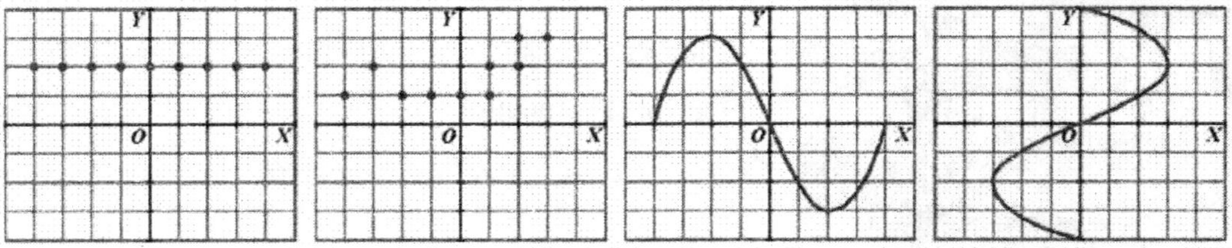

4. Juan has just joined a video-club where he has to pay 6 € as an inscription, 4 € per film (only 10 first) and 2 € per film (for 11th and more).

a) Write a formula or algebraic expression of the function that relates the number of rented films and the cost of them.

b) Build a table with the cost of renting 0, 1, 2, 5, 10, 20, 50 films.

c) Represent the plot of this function.

5. The cost of fuel depending on the kilometers length of a car´s trip is given by the following table:

Distance: x (km)	0	50	100	150	200	250	...
Fuel: f (liters)	0	4	8	12	16	20	...

a) Write the expression of the function $f(x)$, b) Calculate $f(300), f(400)$ y $f(450)$.

Represent the plot of the function. What kind of plot have you obtained?

Exercises

6. We want to build a cylinder shaped well with a radius of 2 m. Express the volume of water (y) that the well is able to contain, depending on its depth (x). Represent its plot.

7. Build a table corresponding to all rectangles with area 36 cm^2 whose base and height are integer numbers. Like this one:

Base: x (cm)							…
Height: y (cm)							…

a) Draw the corresponding plot, representing base in x-axis (*abscise axis*) and height in y-axis (*ordinated axis*).

b) Join the points in a *soft* way.

c) By using the previous table, calculate the perimeters of those rectangles. Draw the plot that relates the base of rectangle and its perimeter, representing the base in the abscise axis and perimeter in the ordinated axis.

4. Graphical representation of functions

Graphs can be given in some different ways:

a) Graphs given by a table.

In a book shop, they have the following table with the cost of photocopies, depending on the number of copies:

Number of copies	5	10	15	20	25	30	…
Price (euros)	2.28	3.37	4.00	5.50	6.12	7.00	…

We can build the corresponding plot by representing these points:

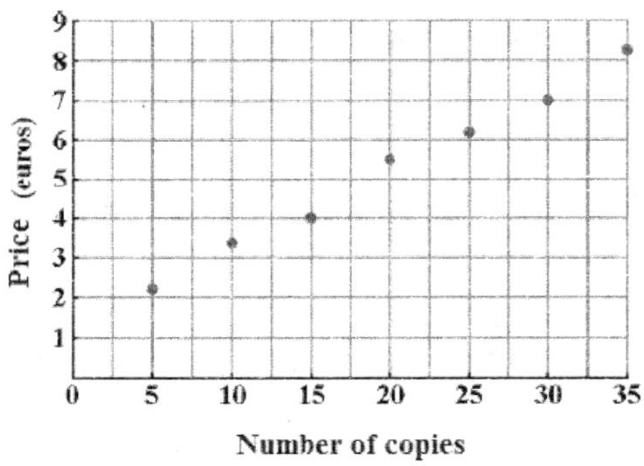

We might have a question: Could I draw intermediate points between these ones? The answer is: not for all, only for a few, because of the nature of magnitudes we are working with. You cannot buy 22.5 copies. You can only by a whole number of copies.

b) Graphs given by a formula.

The expression that gives the area of a circle as a function of its radius is $A = \pi \cdot r^2$. It is a function given by a formula. To represent the function, we create a table, freely introduce values for radius and calculate, by using the formula, corresponding areas.

x =radius	...	...	...	...	...	...	...
y = area	...	...	...	...	...	...	...

We represent the obtained pairs of values and get different points of the plot. If we gave intermediate values to the radius, we would obtain intermediate values of area. Remember in previous example, this was not possible, because you cannot buy 1.5 copies. So, in this case, we can join the points, *in a soft way*.

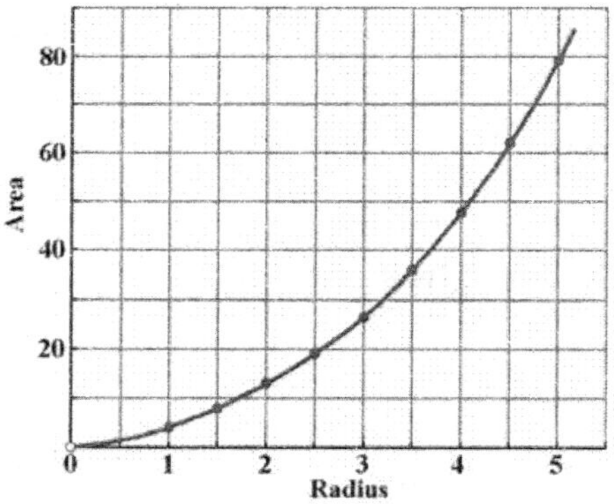

5. Continuous and discontinuous functions

Remember the plots obtained in the previous epigraph:

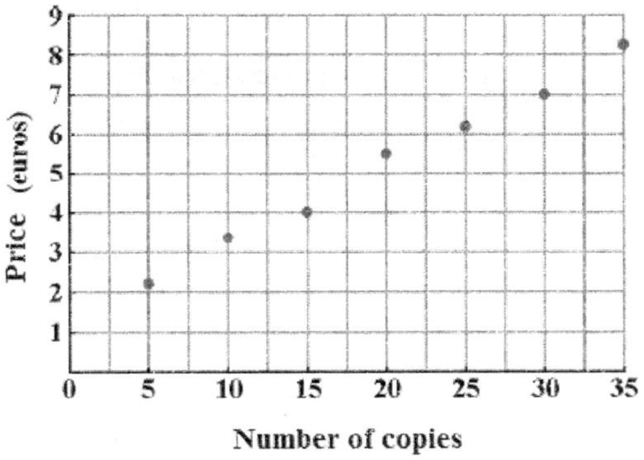

In this first case, independent variable can only take natural numbers and its plot is a series of spots. Functions having this kind of plot are named *discontinuous functions*.

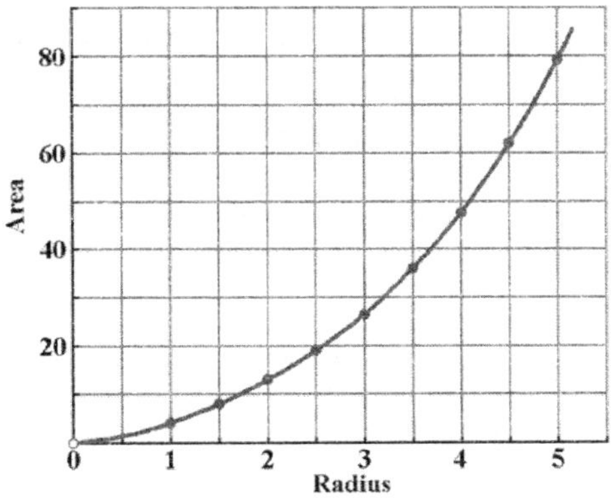

In this other case, independent variable can take all positive real numbers ans its plot is a continuous line. Now we have a *continuous function*.

Imagine now this situation. You are in a telephone box and you are going to phone home. You need 4 coins for the first three minutes. From the third minute, you need a new coin for each three minutes of conversation.

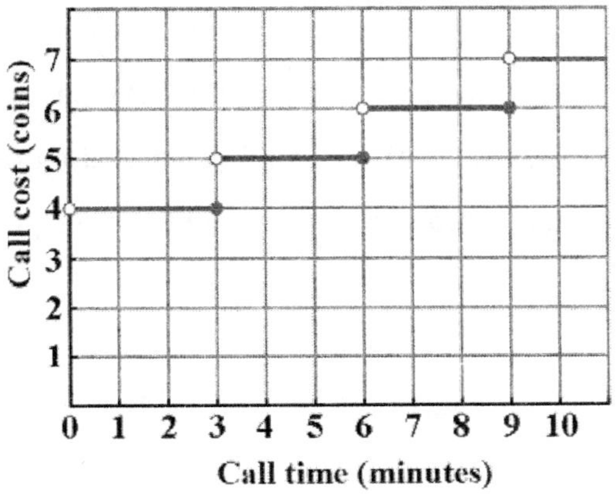

This graph is a discontinuous one. But now, independent variable is continuous (call time), and dependent variable (call cost) shows a series of jumps.

A function $y = f(x)$ is ***continuous*** if its plot can be drawn as an only continuous line. This means, you can draw its plot without picking your pencil up from the paper.

A function $y = f(x)$ is ***discontinuous*** if its plot cannot be drawn as an only continuous line and, therefore, it shows *"jumps"* in its graph. Points where are located these *jumps* are named ***discontinuity points***.

Exercises

8. Study the continuity of the following graphs:

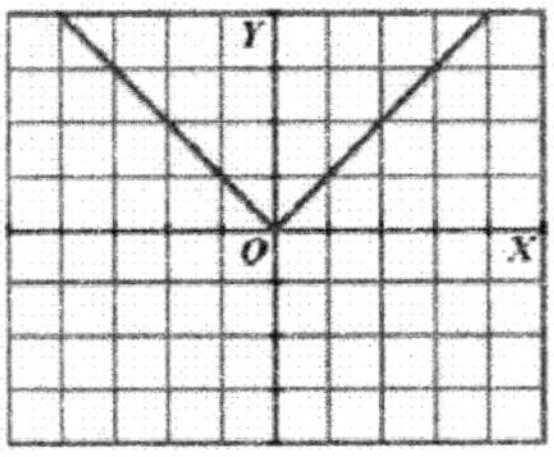

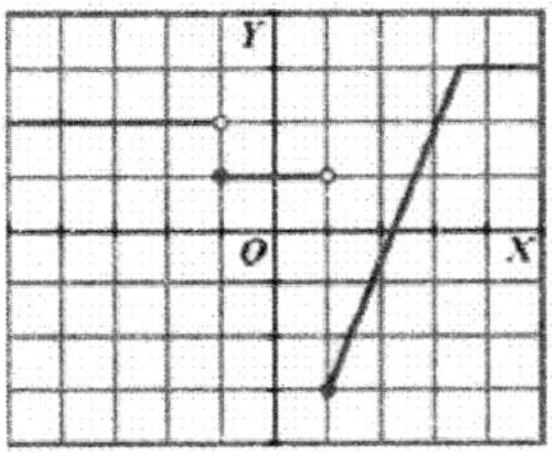

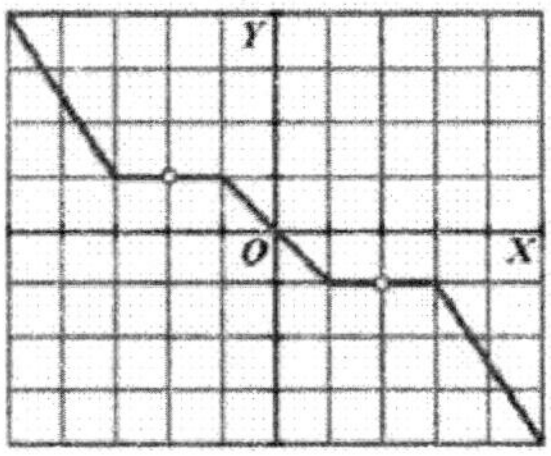

9. Function integer part y = E(x) is defined as the function that relates a real number to its immediately lower integer number. So, for example, E(0.6) = 0; E(2.8) = 2; E(-5.7) = - 6; E(5) = 5. Graphical representation of this function is shown on the right. Study its domain, range and continuity.

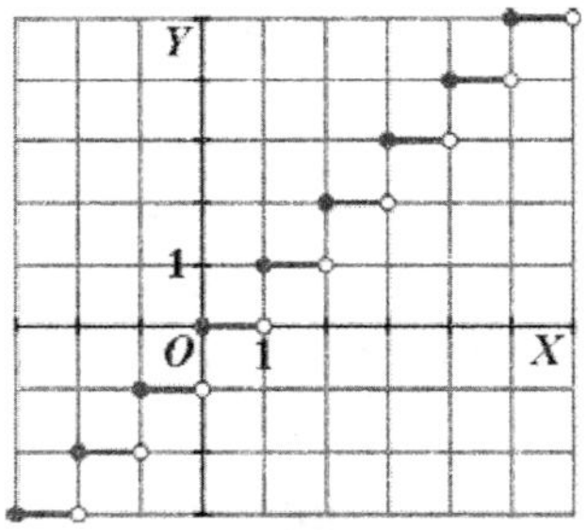

(Exercise 9)

10. The cost of sending a telegram is 1.20 € as a fix tax plus 0.05 € per word. Build a values table with the cost of sending a telegram, depending on the number of words. Is it a continuous function? Why?

11. In a book shop they have different prices for copies of an only original. First 100 have a price of 5 cents/copy, from 101 to 200 copies, 4 cents, from 201 to 500, 3 cents, and more than 500, 2 cents. Represent the function that relates the number of copies with the price per unit.

6. Variations of a function

Introduction: intervals and semi-straight lines

Intervals are segments into the Real Line. There are different types of intervals:

The open interval **(a,b)** includes the real numbers x between a and b, but a and b are NOT included. It is expressed as $a < x < b$.

The closed interval **[a,b]** includes the real numbers x between a and b, being a and b included. It is expressed as $a \leq x \leq b$.

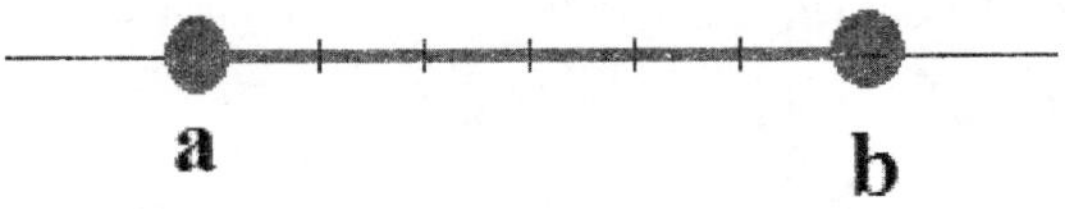

In the same way, the interval **[a,b)** is expressed as $a \leq x < b$, being a included but not b;

while the interval **(a,b]** is expressed as $a < x \leq b$, being b included but not a.

The open semi-straight line **(a, ∞)** includes all the real numbers being higher than a, where a is not included. It is expressed as $a < x$;

while the closed semi-straight line **[a, ∞)** includes all the real numbers being higher than a, being a included. It is expressed as $a \leq x$.

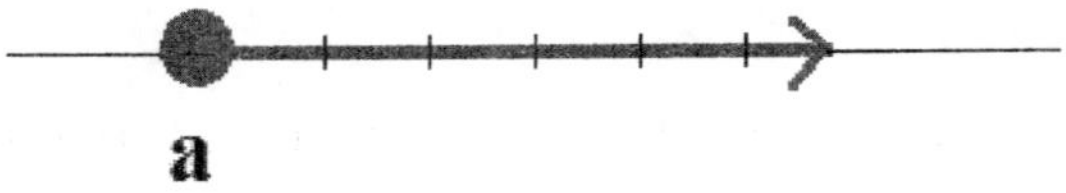

In the same way, semi-straight line **(-∞, b)** includes all the real numbers being lower than b, where b is not included. It is expressed as $x < b$;

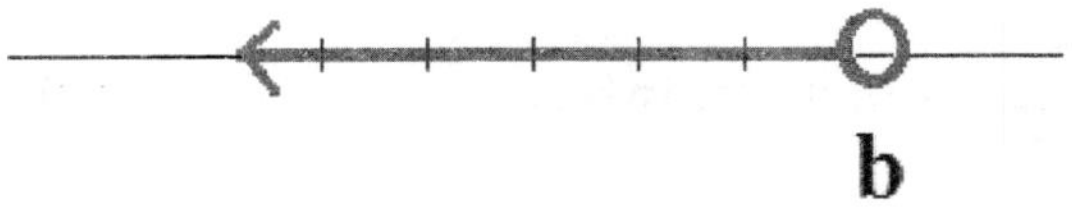

while the semi-straight line **[-∞, b]** includes all the real numbers being lower than b, being b included. It is expressed as $x \leq b$.

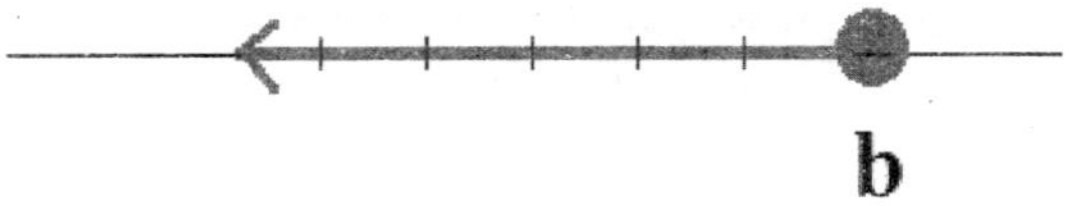

6.1. Increasing and decreasing. Extremes. Maximums and minimums

A function $y = f(x)$ has a ***maximum*** in a point $x = x_0$ if, in values near it, to the left of x_0, function is increasing, and to the right of x_0, function is decreasing. This means that in points near it, function takes values always lower than it.

A function $y = f(x)$ has a ***minimum*** in a point $x = x_0$ if, in values near it, to the left of x_0, function is decreasing, and to the right of x_0, function is increasing. This means that in points near it, function takes values always higher than it.

You are going to see it more clearly in the following figures.

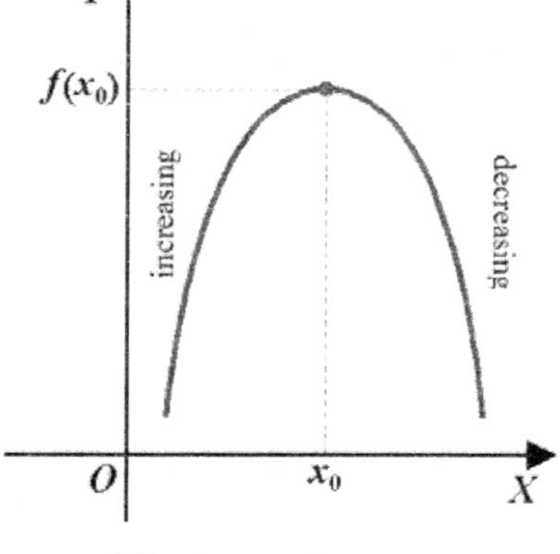

Maximum in $x = x_0$

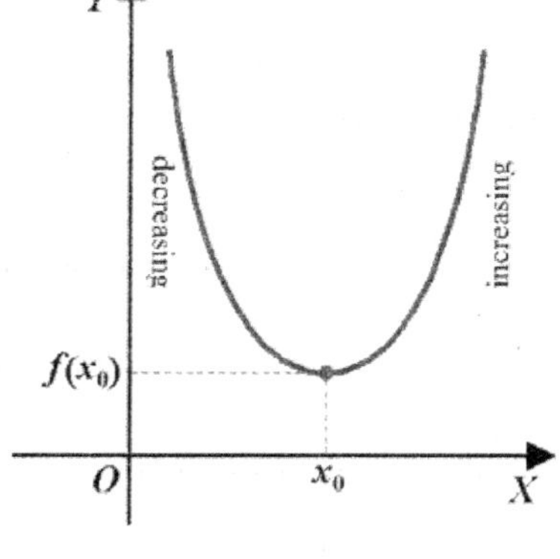

Minimum in $x = x_0$

Anyway, a function can have several minimums and maximums. In order to distinguish them, we define the following concepts.

· A function $y = f(x)$ has an ***absolute maximum (minimum)*** in a point $x = x_0$ if all the values that the function takes are lower (higher) than its image $f(x_0)$.

· A function $y = f(x)$ has a ***relative maximum (minimum)*** in a point $x = x_0$ if near to it, values that the function takes are lower (higher) than its image $f(x_0)$.

Example: Next graph shows the profile of a day at the *"Tour de France"*. Study the increasing and decreasing of the function and its maximums and minimums.

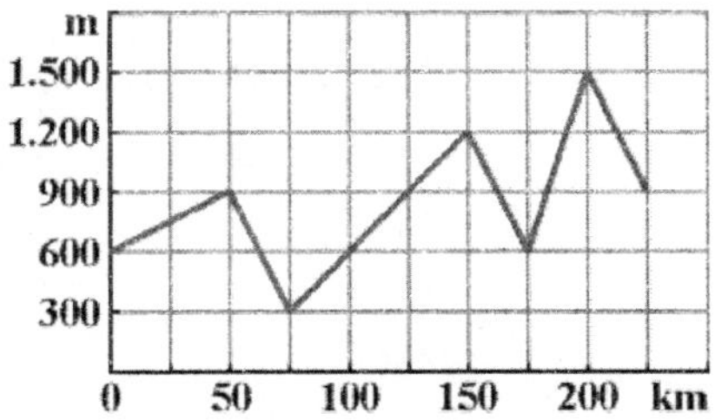

Solution:

· Function is increasing in the intervals [0, 50), (75, 150) and (175, 200).

· Function is decreasing in the intervals (50, 75), (150, 175) and (200, 225].

·It has an absolute maximum at x = 200 km (maximum height is 1500 m) and an absolute minimum at x = 150 km (minimum height is 1200 m).

· Relative maximums are reached at x = 50 km (h = 900 m) and x = 150 km (h = 120 m).

· Relatives minimums are reached at x = 0 km (h = 600 m), x = 175 km (h = 600 m) and x = 225 km (h = 900 m).

Exercises

13. Draw the plot of the function $y = x^2$. By using it, indicate:

a) its increasing and decreasing intervals

b) location of maximums and minimums.

Review exercises

1. Lung capacity measures the volume of air we can expulse form our lungs, and it is measured with an instrument named spirometer. Spirometer registers volume of air that goes into lungs and later goes out of lungs. Graph gives us the volume (liters) into the lungs as a function of time (min).

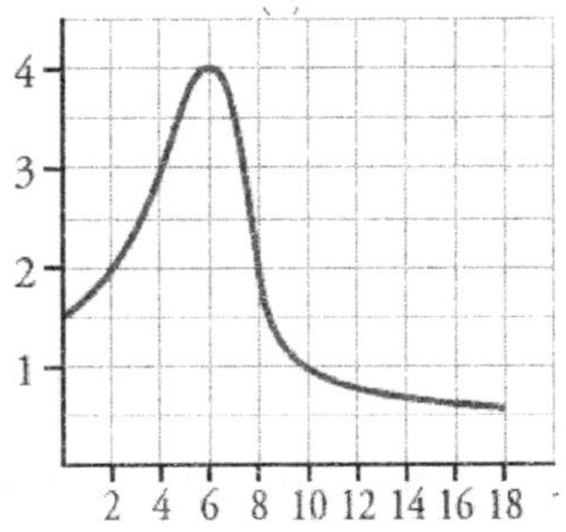

a) What is the initial volume?

b) How long did the observation take?

c) What is the maximum lung capacity of this person?

d) What is its volume 10 minutes after starting? When does it finish?

2. Luis takes 2 hours in going from home to a city 200 km far. There, he has had a business meeting, so, he has stayed at the city for 2 hours. After that, he has gone back home, what has taken him 4 hours.

a) Represent the graph *time-distance to home.*

b) If his speed has been constant during the movement home-city, what is that speed?

c) And in city-home movement?

3. In a city, the want to promote the use of bicycles, so, you can rent a bike. These are prices:

First 3 hours	Free
From third hour	1 €/h

You can rent a bike for a maximum time of 12 hours a day. Draw the plot of the function *time-cost*.

4. Dose of a drug is 0.25 g per kg of the ill person, till a maximum quantity of 15 g.

a) How many grams must have a child whose weight is 10 kg? And another one whose weight is 30 kg? And a 70 kg person?

b) From what weight do people have the same quantity of drug?

c) Represent the function weight of the person-dose.

5. Complete this table, in which are related base and height of rectangles with area 12 m^2:

x = base (m)	1	2	3	4	6	12	x
y = height (m)							

a) Represent the plot of this function.

b) Which of the following is the expression of this function?

$y = \dfrac{x}{12}$ $\qquad$ $y = \dfrac{12}{x}$ $\qquad$ $y = 12x$

6. Join each algebraic expression with their graphical representation:

I) $y = x + 1$ $\qquad$ *II)* $y = x^3$ $\qquad$ *III)* $y = x^2$ $\qquad$ *IV)* $y = -x + 1$

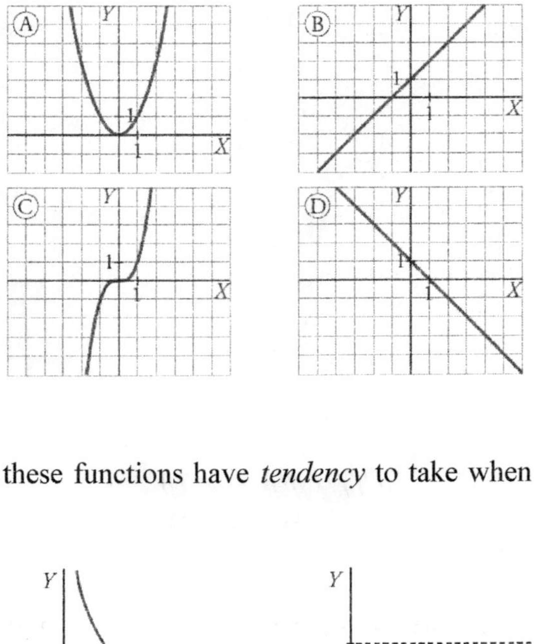

7. Indicate the value that these functions have *tendency* to take when x tends to take very high values.

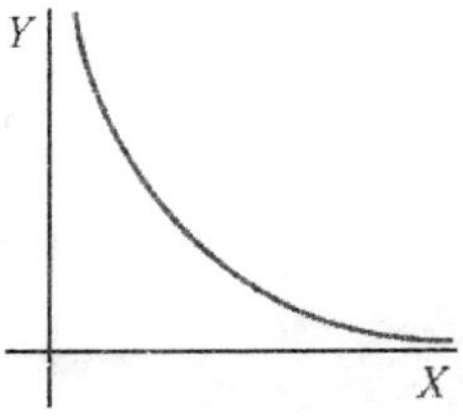

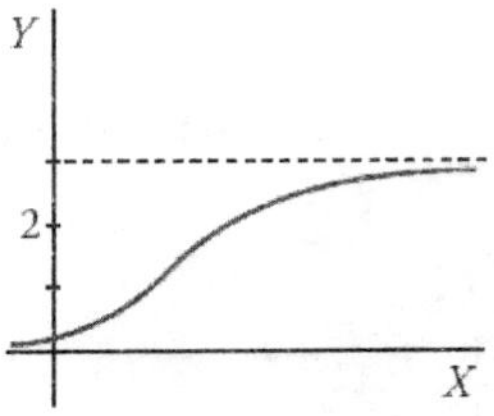

8. Which of the following plots correspond to a function?

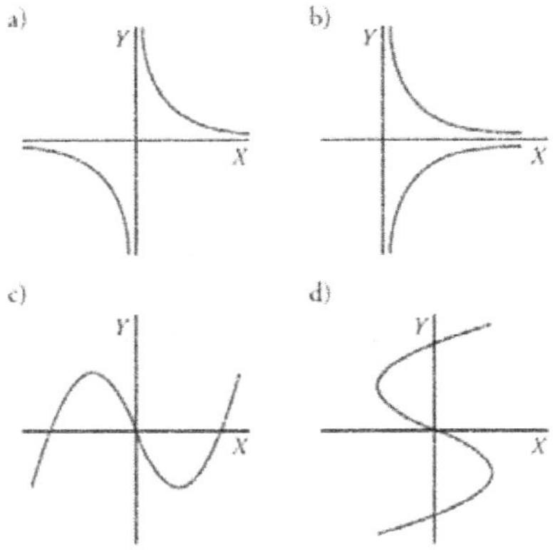

9. For each of the following two graphs, answer to the following questions:

a) Increasing and decreasing intervals.

b) Do they have maximum, minimum or both of them?

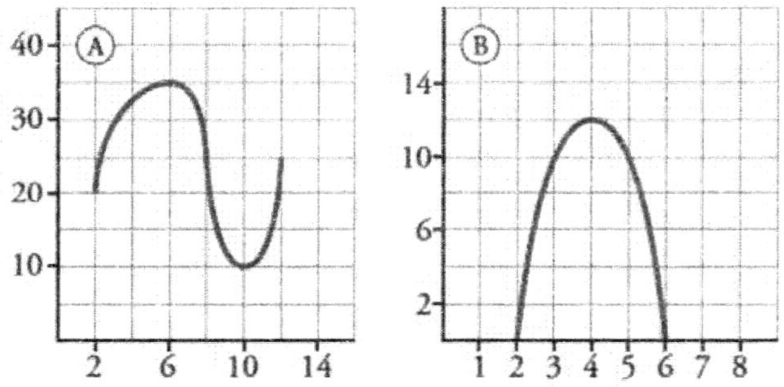

ADVANCED

10. Write, as a function of x, the area of the shadowed zone in each of the following figures:

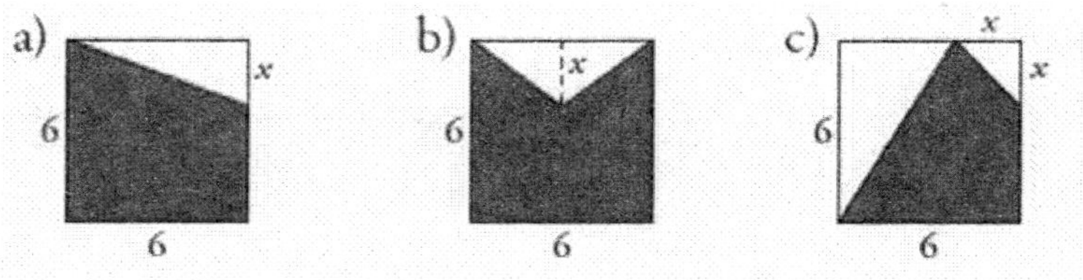

11. This graph shows the variation of the speed of a car when advancing in one of the circuits drawn below.

a) What is the correct circuit?

b) Draw the graph corresponding to the other one.

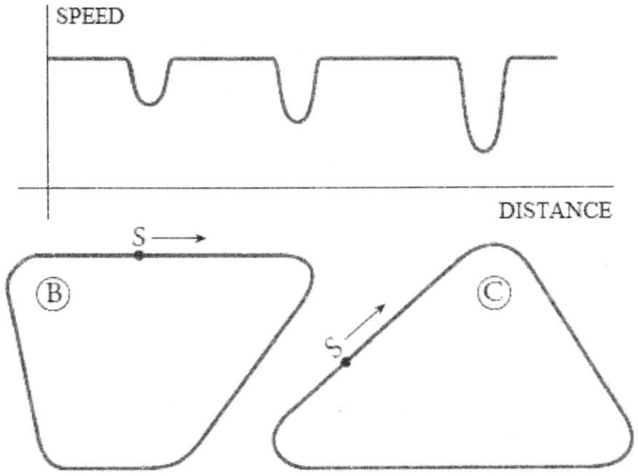

Unit 8.- Linear and quadratic functions

1. Linear function y = mx

AVE train advances with an average speed of 240 km/h. The following table shows advanced distance as a function of trip time.

t = time (h)	1	2	3	4	5	…
e = distance (km)	240	480	720	960	1200	…

As you can see, these magnitudes are in a direct proportion. The algebraic expression of this function is $e = 240t$, where 240 is the proportionality constant. Plot of this function is shown below:

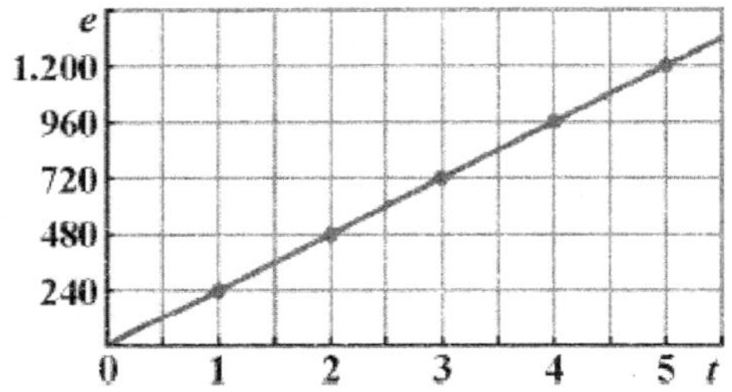

Graph of this function is a continuous ***linear*** plot. This concrete case is an increasing linear plot (distance increases when time increases) and it passes by the origin of the coordinated axes.

Direct proportionality functions or ***linear functions*** are those ones whose graph is a linear plot that passes by the origin of the coordinated axis (0, 0).

Its algebraic expression is $\boldsymbol{y = mx}$, where $\boldsymbol{m}$ is the proportionality constant.

1.2. Slope of a linear plot

In a linear function $\boldsymbol{y = mx}$, coefficient $\boldsymbol{m}$ (proportionality constant) is named ***slope*** and it can be calculated by dividing values of dependent variable by their corresponding values of independent variable.

$$m = \frac{y}{x}$$

Its value indicates the rhythm of increasing or decreasing of the linear plot with equation $\boldsymbol{y = mx}$. It indicates how *y variable* changes when *x variable* changes.

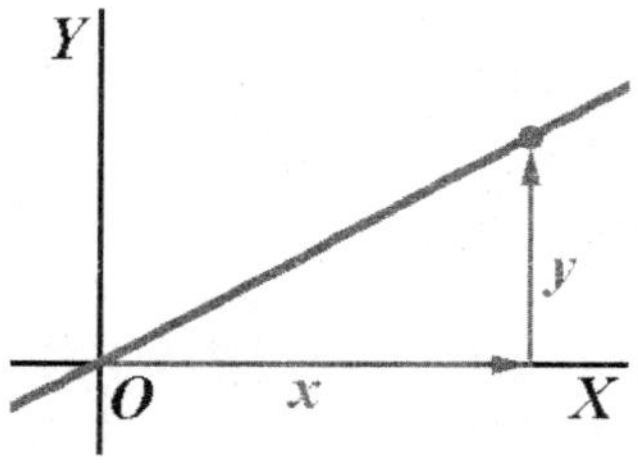

$\boldsymbol{m > 0}$ → linear plot is ***increasing***

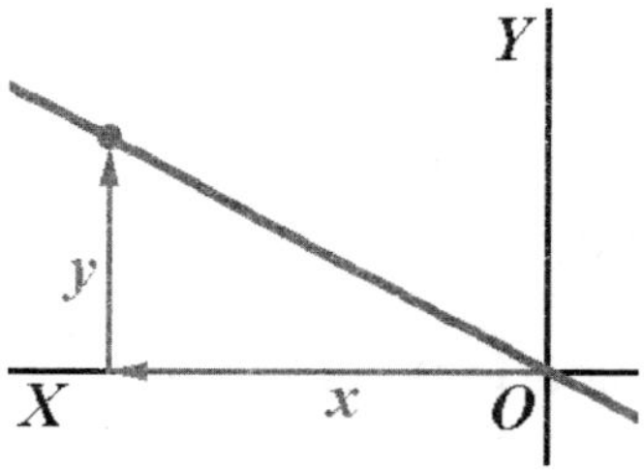

$\boldsymbol{m < 0}$ → linear plot is ***decreasing***

Slope of a linear plot is related to its inclination about the X-axis (angle formed with this axis).

In the following example, you can see that the higher is the slope the higher the inclination of the plot is.

Example: Work out the algebraic expression of the linear functions whose graphical representation is given below:

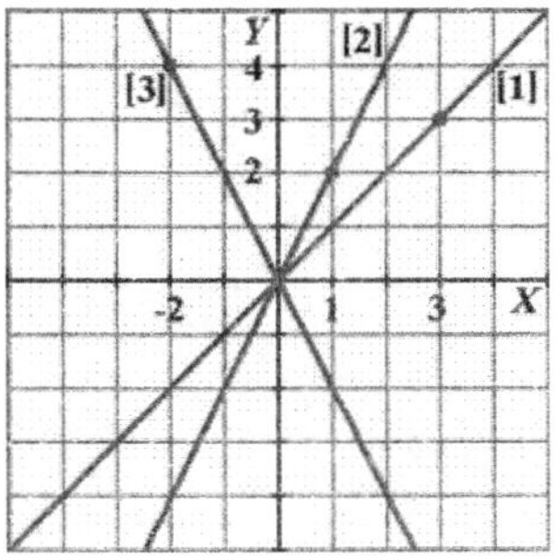

Solution:

Three plots are linear functions with equation $y = mx$, because they are linear plots passing by the origin (0, 0).

Slopes are calculated by working on a point of the plot and calculating the quotient:

(1) $m = \frac{3}{3} = 1$; (2) $m = \frac{4}{2} = 2$; (1) $m = \frac{4}{-2} = -2$;

Linear plots have equations: (1) y = x; (2) y = 2x; (3) y = -2x.

Exercises

1. In a shop, 21 meters of a wire are sold by 3 €.

a) How much is each meter of wire?

b) Build a table indicating the cost of 1, 2, 3, 4, 5, … meters.

c) Draw the corresponding plot and check it is a linear function.

d) Write the algebraic expression of this function. What is the slope or proportionality constant?

2. Following table shows cost and number of photocopies made in a book shop by some students.

x = number of copies	20	40	60	80	…
y = price (euros)	0.80	1.60	2.40	3.20	…

Find the expression that relates the number of copies and their cost. Draw it.

3. Following plots show the relationship existing between volume and mass for some materials.

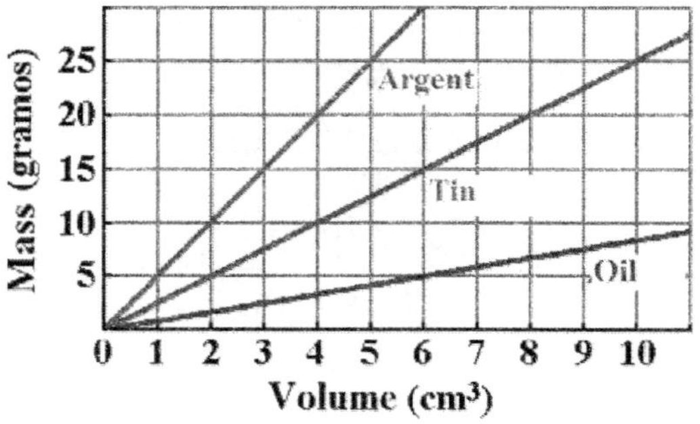

a) Calculate slopes of each linear plot. What is the most dense material? And the least dense? Write down the algebraic expression for all of them.

b) What is the weight of 3 dm^3 of argent?

c) How many liters are occupied by 1 kg of oil?

4. Determine the expression for the functions whose plots are given below.

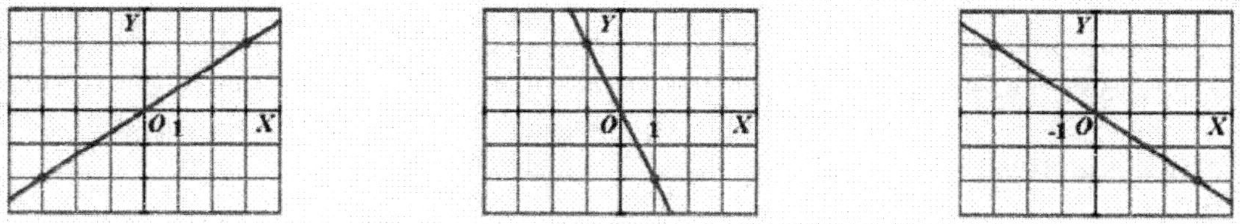

5. Write an equation for each of the following linear functions and draw them on an only coordinated axes:

a) Linear function passing by origin and its slope is 1/2.

b) Linear function passing by origin and by point (-1, 3).

c) Linear function symmetric to $y = 2x$ about the ordinated axis.

d) Linear function symmetric to $y = 2x$ about the abscise axis.

2. Affine function y = mx + n

We have measured temperature of a liquid while it was been heated. Results are shown in the following value table.

x = time (min)	0	1	2	3	4	...
y = temperatura (°C)	15	20	25	30	35	...

Algebraic expression for this function is $y = 5x + 15$ whose graphical representation is shown below.

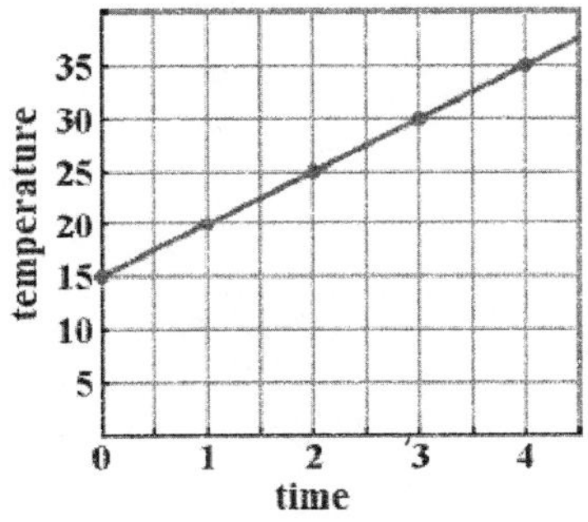

This is not a linear function. Its graph is also a straight line but it does not pass by the coordinate's origin. Its algebraic expression is not $y = mx$, but $y = mx + n$. These functions are named ***affine functions.***

Affine functions are those ones whose graph is a straight line that does not pass by the coordinate´s origin. Its algebraic expression is $y = mx + n$.
In above expression:
· m is the ***slope*** or *gradient* of the function.
· n is the ***y-intercept:*** straight line intersects ordinated axis at point **(0, *n*).**

2.1. Affine functions having the same slope

We have drawn the straight lines corresponding to the following affine and linear functions:

[1] $y = 2x + 4$ [2] $y = 2x$ [3] $y = 2x - 6$

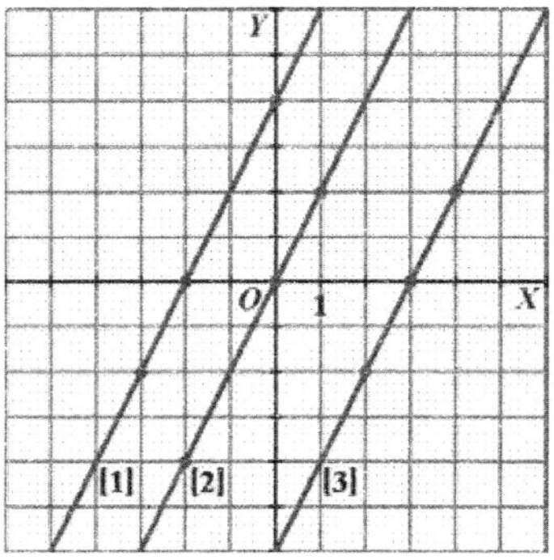

As you can see, they have the same slope, $m = 2$, and their corresponding plots are *parallel* straight lines.

Plots of affine linear functions having ***the same slope m*** are ***parallel straight lines.***

If they have ***different slopes***, they will be ***secant*** straight lines.

2.2. Affine functions having the same *y-intercept*

We are representing following affine functions:

[1] $y = x + 4$ [2] $y = 2x + 4$ [3] $y = -x + 4$

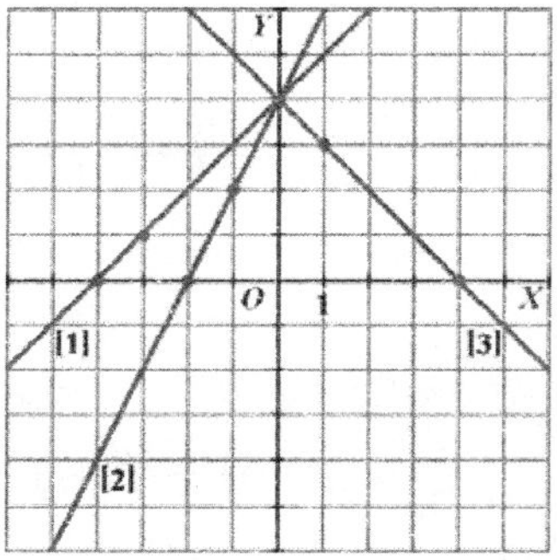

As you can see, all of them have 4 as their *y-intercept*, $n = 4$. This makes all straight lines pass by the point (0, 4).

Graphs of affine functions having ***the same y-intercept n*** are ***secant straight lines*** that intersect in the point **(0, *n*).**

Example: Given straight lines with equations $y = 2x + 5$ and $y = 5x + 2$, are they parallel or secant? Draw them at the same coordinated axes:

<u>Solution:</u> As they have different slopes, they are secant. If we draw them, we can see they intersect at point (1, 7).

x	$y = 2x + 5$
−1	3
0	5
1	7
...	...

x	$y = 5x + 2$
−1	−3
0	2
1	7
...	...

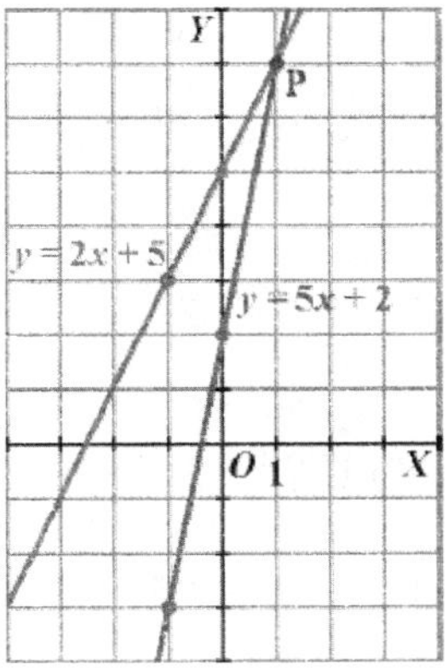

Exercises

6. A windows manufacturer sells his windows with a price of 3 €/m of the frame and 12 € for glass (independently on the area).

a) How much does it cost a square window with side 2 m?

b) Price for a window has been 60 €, how long is its side?

c) Find the expression that gives us the price of a window as a function of the dimensions and draw it.

7. Cost of electrical energy in a house is given by the contracted power, 12 €, and the price of each consumed kilowatt·h (kw·h), that costs 0.15 €/kw·h.

a) What is the expression that gives the cost, as a function of consume? Draw it.

b) How much has to pay a family whose consume has been 200 kw·h?

8. RENFE has special prices for students who want to travel through Europe in summer. A student must pay a fix tax of 30 € plus 0.02 €/km.

a) Write the equation that relates the cost with the distance, indicating which are independent and dependent variables.

b) Draw the plot of the function.

c) Calculate how much must pay a student for a trip through France for 5400 kilometers.

d) How many kilometers has a student travelled if he has paid 94 €?

Exercises

9. Following table belongs to an affine function $y = mx + n$.

x	0	10	20	30	40	50
y	−3		37			97

a) Complete the table and represent its graph.

b) Find its algebraic expression, indicating its *slope* and *y-intercept*.

10. Among following straight lines, indicate which are parallel and which are secant.

a) $y = 3x + 2$ b) $y = 2x + 3$ c) $y = 3x - 3$ d) $y = 2x - 1$ e) $y = 3x - 1$

For those being parallel, draw them on the same coordinated axes.

3. Equations of a straight line

3.1. Slope of a straight line passing by 2 points

Given two points with coordinates $A(x_1, y_1)$ and $B(x_2, y_2)$, we can calculate the slope of the straight line that joins them. We have only to calculate variation of *y-variable* (increasing or decreasing) when

x- variable varies, passing from point A to B.

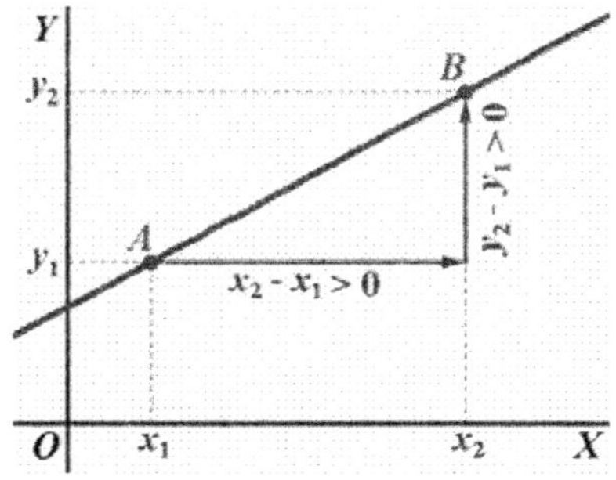

Slope, m, of a straight line passing by points $A(x_1, y_1)$ and $B(x_2, y_2)$ is calculated as

$$m = \frac{y_2 - y_1}{x_2 - x_1}$$

Example: Calculate the slope of the straight line passing by points with coordinates $A(-1, 1)$ and $B(1, 5)$.

<u>Solution:</u>

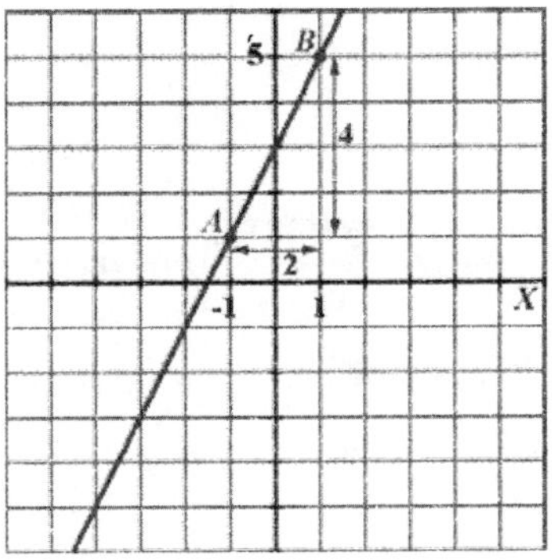

$$m = \frac{y_2 - y_1}{x_2 - x_1} = \frac{5-1}{1-(-1)} = \frac{4}{2} = 2$$

3.2. Equation of the straight line passing by 2 points

There are two methods to calculate the equation for the straight line passing by 2 points.

Example: Write the equation for the straight line passing by points with coordinates $A(-1, 1)$ and $B(1, 5)$.

<u>Method 1:</u>

We calculate the slope: $m = \frac{y_2 - y_1}{x_2 - x_1} = \frac{5-1}{1-(-1)} = \frac{4}{2} = 2$

Affine function equation is: $y = mx + n$. So, $y = 2x + n$. We only need to know n value.

We freely choose one of the points by which this function passes. For example: $B(1, 5)$.

As function passes by B, its coordinates need to verify equation of this function. We have only to substitute: $y = 2x + n \rightarrow \left| \begin{matrix} x = 1 \\ y = 5 \end{matrix} \right| \rightarrow 5 = 2 \cdot 1 + n \rightarrow 5 = 2 + n \rightarrow n = 3$

So our equation is $\boxed{y = 2x + 3}$.

<u>Method 2:</u>

As function is passing by points A and B, their coordinates need to verify equation $y = mx + n$

If we substitute coordinates of points A and B in this equation, we will have a system of linear equations.

A(-1,1): $y = mx + n \rightarrow \begin{vmatrix} x = -1 \\ y = 1 \end{vmatrix} \rightarrow 1 = m\cdot(-1) + n \rightarrow 1 = -m + n \rightarrow -m + n = 1$

B(1,5): $y = mx + n \rightarrow \begin{vmatrix} x = 1 \\ y = 5 \end{vmatrix} \rightarrow 5 = m\cdot 1 + n \rightarrow 5 = m + n \rightarrow m + n = 5$

We have the following system: $\begin{cases} -m + n = 1 \\ m + n = 5 \end{cases}$ Its solution is (m = 2, n = 3).

So our equation is $\boxed{y = 2x + 3}$.

3.3. Equation of the straight line passing by points, knowing its slope or y-intercept

We are going to see it with some examples:

Example: Find the equation of the straight line being parallel to the one with equation $y = 4x - 2$, knowing it passes by the point with coordinates (1, 9).

<u>Solution:</u>

The straight line we are looking for is parallel to $y = 4x - 2$. So, its slope is $m = 4$, and its equation will be $y = 4x + n$.

As it passes by point (1, 9), this coordinates must verify $9 = 4\cdot 1 + n$, so $n = 9 - 4 = 5$.
Our straight line is $\boxed{y = 4x + 5}$.

Exercises

11. Determine expressions for functions whose plots are given below:

y = mx+n   Si pasa por 0-0 : y=mx

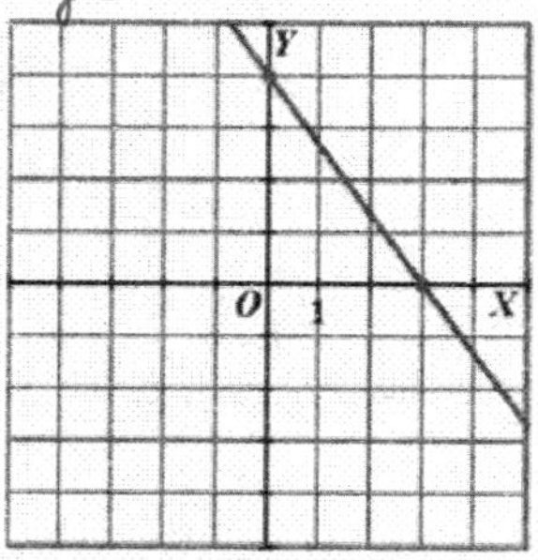

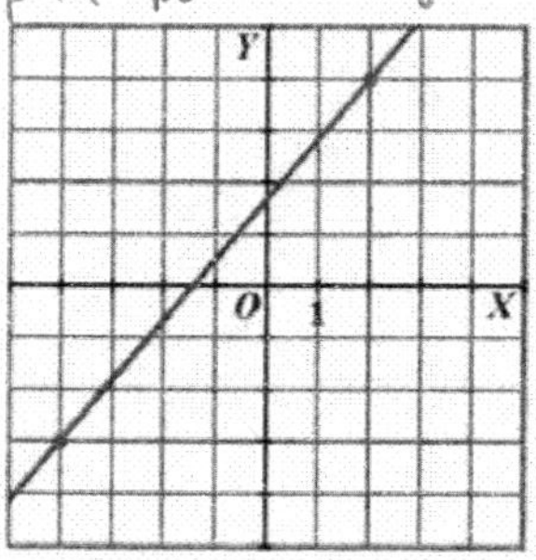

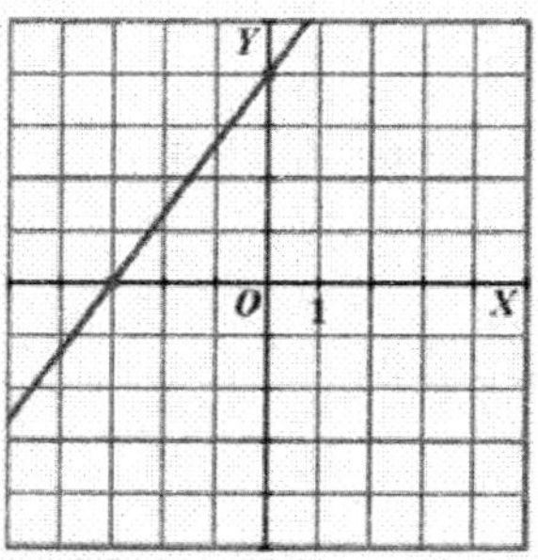

(0,4) n=4

$m = \frac{-4}{3} < 0$ $y = \frac{-4}{3}x + 4$

Exercises

12. Find the equations of the following straight lines and draw them on an only coordinated axes:

a) Straight line whose slope is 3 and y-intercept is 2.

b) Straight line whose slope is 2 and passes by point (2, 7).

c) Straight line being parallel to straight line with equation $y = 3x - 5$ and passing by the point (-2, 3).

d) Straight line passing by the point (3, 0) and its y-intercept is 3.

e) Straight line passing by the points with coordinates $A(-3, 5)$ y $B(1, 5)$.

13. Given straight line $y = 2x + 4$, find equations for the following straight lines (it might be useful a graphical scheme)

a) Symmetric straight line to the given one, about *Y-axis*.

b) Symmetric straight line to the given one, about *X-axis*.

c) Symmetric straight line to the given one, about *coordinate's origin*.

4. Straight lines being parallel to the coordinated axes

4.1. Straight lines being parallel to the *X- axis*

If you think on a horizontal line, you know it has a null slope.

If $m = 0$, affine function $y = mx + n$ becomes into $y = n$, which is the algebraic expression of a *constant function*. Its plot is parallel to X-axis passing by the point with y-intercept $y = n$.

Example: Graph for function $y = 2$ is a straight line being parallel to X-axis, passing by point (0, 2). Notice all points of the line have ordinated 2, having the form $(x, 2)$.

x	$y = 2$
−2	2
−1	2
0	2
1	2
2	2
...	...

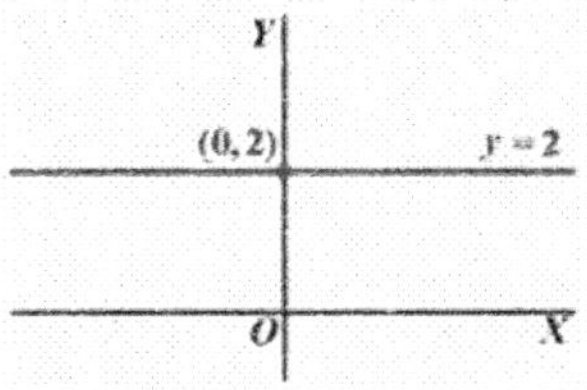

A ***constant function*** is that one in which value for dependent variable does not vary, it is always the same.
Its plot is a ***straight line*** being ***parallel to X-axis*** and its expression is $y = n$.

4.2. Lines being parallel to the *Y- axis*

In the following straight line, parallel to *Y-axis*, we can observe that all points on it, have different values of *y*, but all of them have the same abscise value, x = 2. So, its graph **is not a function because** to an only value of *x, (x = 2),* corresponds infinite values of *y*.

Some points of the line are (2, -1), (2, 0), (2, 2), ... They have in common their abscise, *x = 2.* That is the reason by which its equation is $x = 2$.

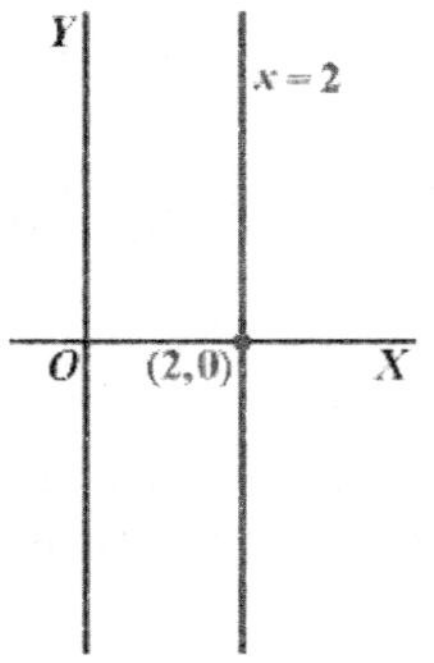

Straight lines being **parallel to the Y-axis** passing by the point
(*a*, 0) have an equation $x = a$.

These straight lines **are not graphs of functions.**

Exercises

14. Represent the straight lines with equations $y = 2$, $y = 3$, $x = -2$, $x = 2$.

15. Find the equation of the following straight lines and draw them on the same coordinated axes:

a) Straight line being parallel to the abscise axis and passing by the point (4, 3).

b) Straight line being parallel to the ordinated axis and passing by the point (-1, 7).

16. Find the equations of the following straight lines:

a) Having slope 3 and y-intercept -7.

b) Being parallel to the straight line with equation $x = 5$ and passing by the point (-2, 8).

c) Having slope 5 and passing by the point (-1, -2).

d) Passing by the points (2, 3) and (-1, 6).

e) Intersecting the *Y-axis* at the point -3 and passing by the point (-2, 1).

f) Passing by point (2, -1) and being parallel to the straight line with equation $y = 6x + 9$.

g) Being parallel to the *X-axis* and passing by the point (-3, 2).

17. Locate intersection with axes for the following straight lines. When you have found the, draw them on a coordinate´s system and draw the straight lines.

a) $y = 3x - 6$ b) $y = 2x + 5$ c) $y = 3x$ d) $y = -3$ e) $x = 4$

5. Quadratic functions $y = ax^2 + bx + c$

Exercises

18. Calculate the values of a, b and c, for the following quadratic functions:

a) $y = 3x^2 - x + 5$ b) $y = 5x - x^2$ c) $y = x^2 + x - 8$

d) $2y = 6x - x^2 + 4$

19. Given the quadratic function $y = x^2 - 2x - 3$, represent its curve by calculating the images of $x = -1, 0, 1, 2$ and 3. What kind of graph have you obtained?

Quadratic functions are those whose algebraic expression is $y = ax^2 + bx + c$. Its graph is a parable, whose vertex is located at the abscise: $x_v = \frac{-b}{2a}$

IMPORTANT: The parable will be **open up** if a > 0 and **open down** if a < 0.

Exercises

20. Find out the vertex of the following parables:

a) $y = 3x^2 - 12x + 1$ b) $2y - 3x + x^2 = 5 - x - x^2$

Remember one point has two coordinates (x_v, y_v).

21. Represent the curve of the following parables, finding out the vertex, some close points and the intersection points with the axes:

a) $y = (x + 4)^2$ b) $y = x^2 + 2x$ c) $y = -3x^2 + 6x - 3$ d) $y = -x^2 + 5$

22. Join each quadratic function to its curve:

a) $y = x^2$

b) $y = (x - 3)^2$

c) $y = x^2 - 3$

d) $y = x^2 - 6x + 6$

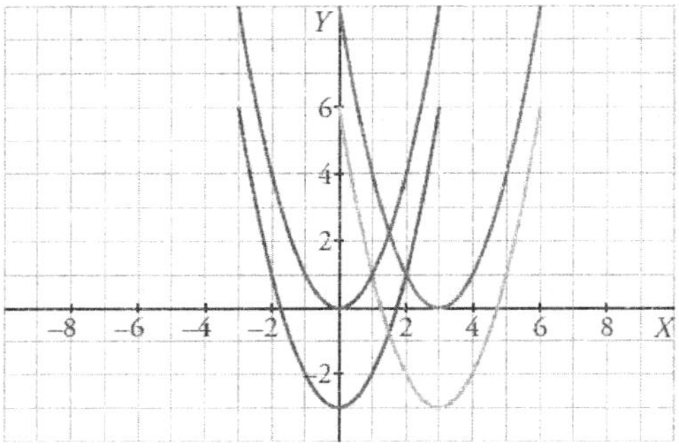

23. Locate the vertex (abscise and ordinated) of the following parables and indicate if it is a maximum or a minimum:

a) $y = x^2 - 5$ b) $y = 3 - x^2$ c) $y = -2x^2 - 4x + 3$

d) $y = 3x^2 - 6x$ e) $y = 5x^2 + 20x + 20$ f) $y = -x^2 + 5x - \frac{3}{2}$

24. Jason jumped off of a cliff into the ocean in Acapulco while vacationing with some friends. His height as a function of time could be modeled by the function $h(t) = -16t^2 + 16t + 480$, where t is the time in seconds and h is the height in feet.

a) How long did it take for Jason to reach his maximum height?

b) What was the highest point that Jason reached?

c) Jason hit the water after how many seconds?

Exercises

25. You are trying to dunk a basket ball. You need to jump 2.5 ft. in the air to dunk the ball. The height that your feet are above the ground is given by the function $h(t) = -16t^2 + 12t$. What is the máximum height your feet will be above the ground? Will you be able to dunk the basketball?

26. One of the games at a carnival involves trying to ring a bell with a ball by hitting a lever that propels the ball into the air. The height of the ball is modeled by the equation $h(t) = -16t^2 + 39t$. If the bell is 25 ft. above the ground, will it be hit by the ball?

Review exercises

1. Represent the following straight lines:

a) y = 4x; b) y = -3x; c) $y = -\frac{x}{2}$ d) y = -4.

2. Represent these straight lines:

a) y = 0.6x; b) $y = \frac{1}{2}x$; c) $y = -2.4x$ d) $y = -\frac{2}{5}x$.

3. Represent these straight lines:

a) y = 15x; b) $y = -25x$; c) $y = \frac{x}{200}$ d) $y = -\frac{1}{120}x$.

4. Represent these straight lines:

a) y = -2x + 1; b) $y = -\frac{x}{2} + 3$; c) $y = -\frac{8}{5}$ d) $y = \frac{3x-5}{2}$.

e) y = 2.5x - 1; f) $y = \frac{3}{4}x + \frac{1}{2}$;

5. Represent the straight line 3x – 2y + 4 = 0

6. Represent these straight lines:

a) $x + y = 5$ b) $2x - y = -3$ c) $2x - 3y = 12$ d) $3x + 2y = -6$

e) $4x + 9y = 0$ f) $4x - 5y + 20 = 0$

7. Represent, on the same coordinated axes, each pair of straight lines, and determine their intersection:

a) $\begin{cases} 2x+3y=7 \\ y=-x+3 \end{cases}$ b) $\begin{cases} y=-4x+1 \\ y=3 \end{cases}$ c) $\begin{cases} y=1-3(x+2) \\ 2x+y+3=0 \end{cases}$ d) $\begin{cases} x-y=2 \\ 3x+y=2 \end{cases}$

8. Find the slope and write the equations for these straight lines:

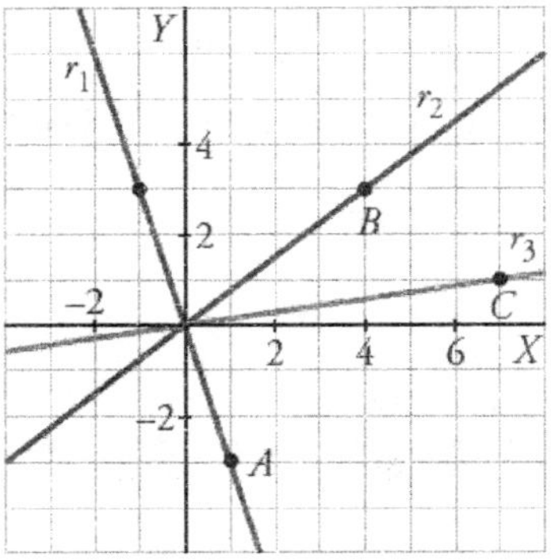

9. Write the equation for the straight line passing by the origin and by the point:

a) P (12, –3) b) P (–2, 3/4) c) P (–7, –21) d) P (30, 63)

10. Find the equation of the linear function passing by point (-5, 25) that is parallel to X-axis.

11. Write equations, *in the point-slope form*, for straight lines knowing their slope and a point:

a) P(-2, 5), m = 3; b) P(1, -5), m = - 2; c) P(-7, 2), m = 3/2; d) P(-2, -4), m = -2/3.

12. Write the equations of provious exercise in the general form.

13. Write the equation of the represented functions. Which are increasing and which decreasing? Think on the sign of their slope.

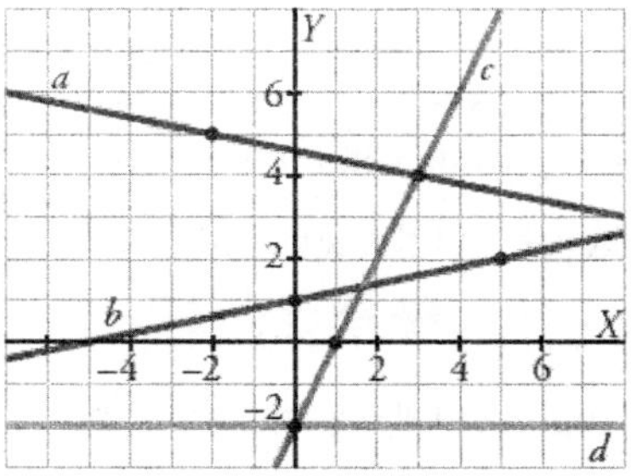

14. Calculate the slope of the straight lines passing by each pair of points and write their equations:

a) A(2, –1), B(3, 4) b) A(–5, 2), B(–3, 1) c) A(–7, –2), B(9, –3)

d) A(0, 6), B(–3, 0) e) A(3/2, 2), B(1, 2/3) f) A(-1/2, 3/4), B(1/3, 1)

15. Join each straight line, r, s, t, p, and q, and its equation:

a) $y = -\frac{1}{3}x$ b) $y = \frac{3}{2}x + 1$

c) $y = \frac{2}{5}x$ d) $y = \frac{2}{5}x + 2$

e) $y = -2$

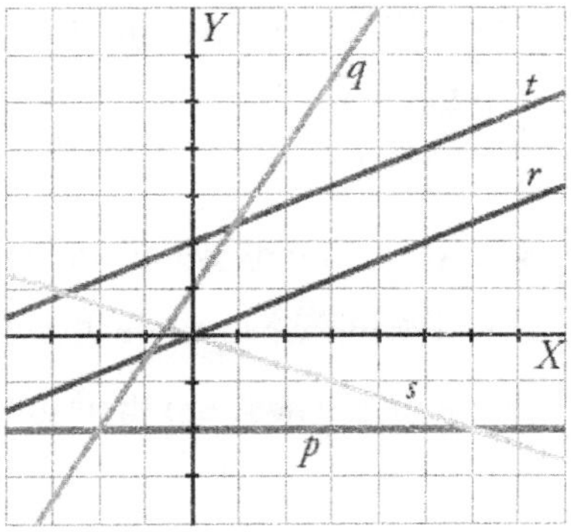

16. Indicate the slope of the following straight lines. After that, draw them on an only coordinated axes and observe the relationship existing between them. Do you get any conclussion?

a) $y = 2x$ b) $y = 2x - 3$ c) $2x - y + 1 = 0$ d) $4x - 2y + 5 = 0$

17. Write the equation of all these straight lines and draw them:

a) Passing by $(-3, 2)$ and $(1, -4)$. b) Passing by (2/5, -1) and slope -1/2.

c) Passing by (2, 1) and its y-intercept is 3. d) Passing by $(2, -4)$ and being parallel to $y = 3x$.

e) Parallel to X-axis and passing by $(-2, -4)$.f) Parallel to Y-axis and passing by $(-2, -4)$.

18. Find out the equation of the straight line that passes by the point $(2, -1)$ and it is parallel to another one passing by the points (3, 0) y (2, 5).

19. Check that the point (23, 74) belongs to straight line $y = 4x - 18$.

20. Calculate the value of c so that straight line $3x - 5y = c$ passes by the point $(-2, 4)$.

21. Calculate the value of b so that straight line $2x + by = -11$ passes by the point $(2, -5)$.

22. Calculate the slope and y-intercept of the following straight lines:

a) $-5x + 8y = 3$ b) $4x - 7y = -8$ c) $3y = 12$ d) $6x - 2y - 3 = 0$

23. Check if there is a straight line passing by the three points $A(1, 1)$ $B(-1, -2)$ $C(65, 97)$.

24. In an English school you must pay an initial quantity of 10 € plus 15 €/month.

a) Find the expression of the function *number of months* → *Total cost.*

b) Draw it.

25. Two artificial water wells, *A* and *B*, work in the following way: while A well is being filled, B is emptying. These are their graphs:

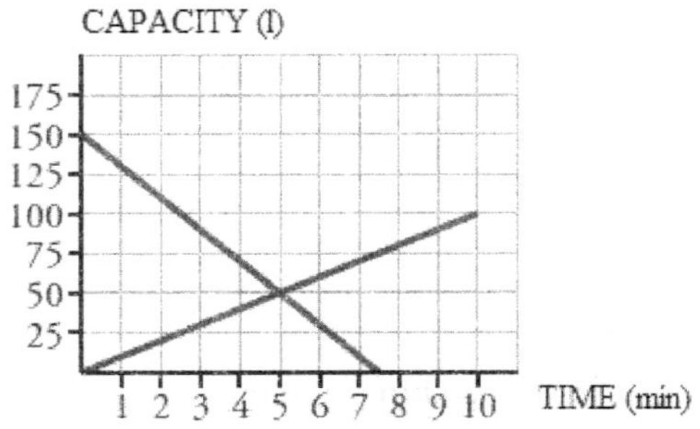

a) Indicate which is the graph corresponding to A and which to B.

b) What are the filling and emptying speeds?

c) When do both wells have the same quantity of water?

26. In a swimming pool there are 200 m^3 of water. When opening the tap, it empties with a constant rhythm of 40 m^3/min.

a) How long does it take to get empty?

b) Write the algebraic expression of the function that relates quantity of water in the swimming pool, with time.

c) Draw it and calculate its domain.

27. In function $y = mx + n$, how must the value of m be to make it decreasing?

28. Represent each straight line and indicate if they are a function or not:

a) $y = 5$ b) $x = -2$ c) $3y + 2 = 0$ d) $x - 4 = 0$

29. Consider the straight line $y = \frac{5}{2}x + 4$.

a) Write equations for two straight lines being parallel to it.

b) Write the equation of a straight line with the same y-intercept.

30. What is the slope of the straight line y = - 7?

31. What is the straight line with equation $y = 0$? And $x = 0$?

32. Write the equation of vertical straight line passing by (3,5)

33. Consider the following straight lines. Calculate their slopes and, without drawing them, indicate which of them are parallel:

a) $y = 5x - 1$ b) $5x - y + 3 = 0$ c) $y = -5x + 1$ d) $y = \frac{5x-1}{2}$

34. Indicate if the following statements are true or false:

a) Straight line $x = 5$ is parallel to abscise axis.

b) Straight line $x - 2 = 0$ is parallel to ordinated axis.

c) Straight line $y = -4$ is parallel to abscise axis.

d) Straight lines $y = 3x - 2$ and $y = 2x - 3$ are parallel.

35. Without drawing them, calculate the intersection points with the axes for the following straight lines:

a) $x - y = 4$ b) $3x - y = 6$ c) $y = \frac{x-2}{4}$ d) $y = -\frac{2}{3}x + 1$

36. Represent the curve of the following parables, finding out the vertex, some close points and the intersection points with the axes:

a) $y = x^2 + 6x + 1$ b) $y = x^2 - 2x$ c) $y = 3x^2 - 6x + 3$ d) $y = x^2 - 5$

37. Locate the vertex (abscise and ordinated) of the following parables and indicate if it is a maximum or a minimum:

a) $y = x^2 + 5$ b) $y = 3 + 2x^2$ c) $y = 2x^2 - 6x + 4$
d) $y = 3x^2 + 6x$ e) $y = 5x^2 - 20x + 20$ f) $y = x^2 - 5x - 4$

38. An amateur rocketry club is holding a competition. There is cloud cover at 1000 ft. If a rocket is launched with a velocity of 315 ft/s, use the function $h(t) = -16t^2 + vt + h_0$ to determine how long the rocket is out of sight.

39. A ship drops anchor in a harbor. The anchor is 49 ft. above the surface of the water when it is released. Use the vertical motion formula $h(t) = -16t^2 + vt + h_0$ to answer the following questions.

a) What is the value of h_0, the starting height?

b) What is the value of h when the anchor hits the water?

c) The starting velocity is zero. After how many seconds will the anchor hit the water?

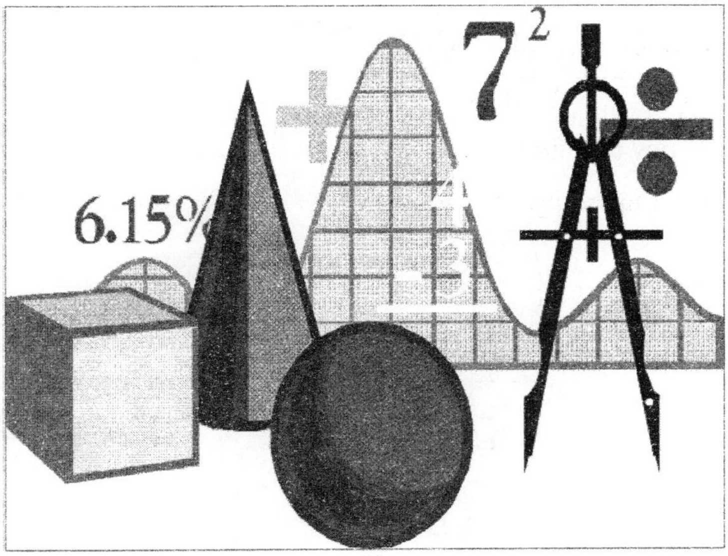

Unit 9.- Metric problems in plane geometry

1. Angles in straight lines

An **angle** is a shape, formed by two lines or rays diverging from a common point (the *vertex*).

Parts of an angle:

Vertex The vertex is the common point at which the two lines or rays are joined.

Legs The legs (sides) of an angle are the two lines that make it up.

Interior The interior of an angle is the space between the sides of the angle, extending out to infinity.

Exterior All the space on the plane that is not the interior.

Angles are measured in ***degrees (º)***. For example, a whole round is equal to 360º.

Bisector of an angle

Given an angle, its **bisector** is the semi-straight line that, passing by its vertex, divides the angle into two identical angles.

Types of angles

There are six types of angles:

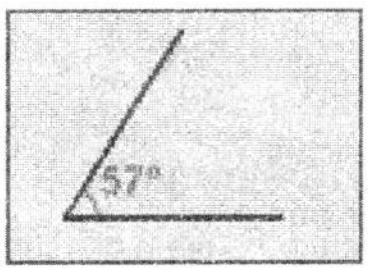

Acute angle
Less than 90°

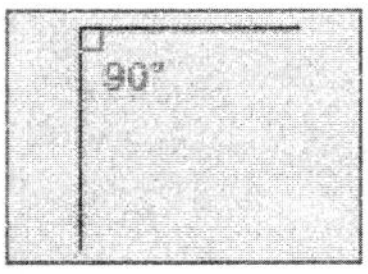

Right angle
Exactly 90°

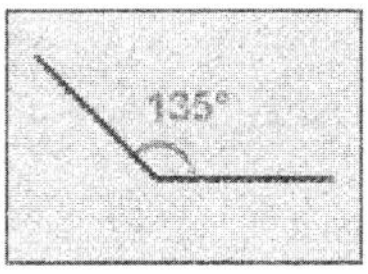

Obtuse angle
Between 90° and 180°

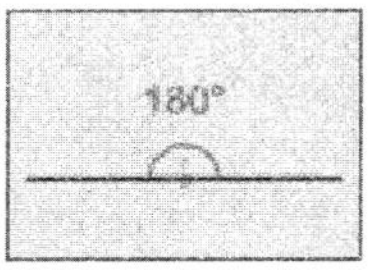

Straight angle
Exactly 180°

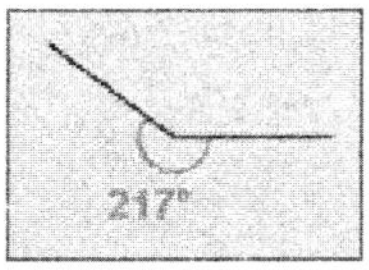

Reflex angle
Between 180° and 360°

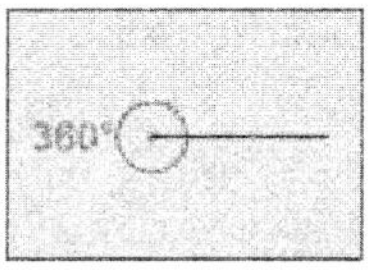

Full angle
Exactly 360°

Angles relationships

Complementary angles: Two angles that add up to 90°.

Supplementary: Two angles that add up to 180°.

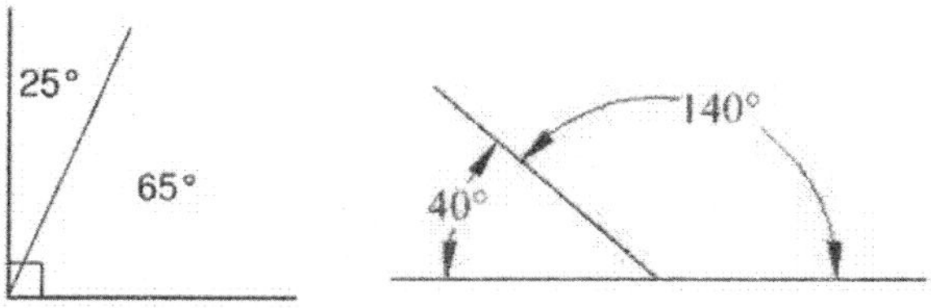

Complementary and supplementary angles

Consecutive angles: Two angles that share a common side and a common vertex, but do not overlap.

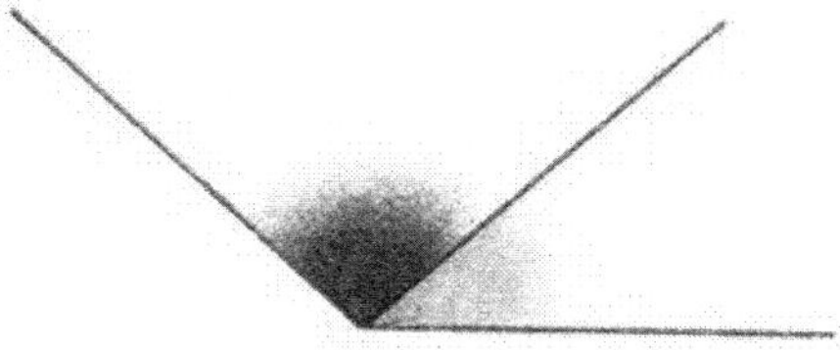

Adjacent angles: Two angles that share a common side and vertex, being the other two sides on the same straight line.

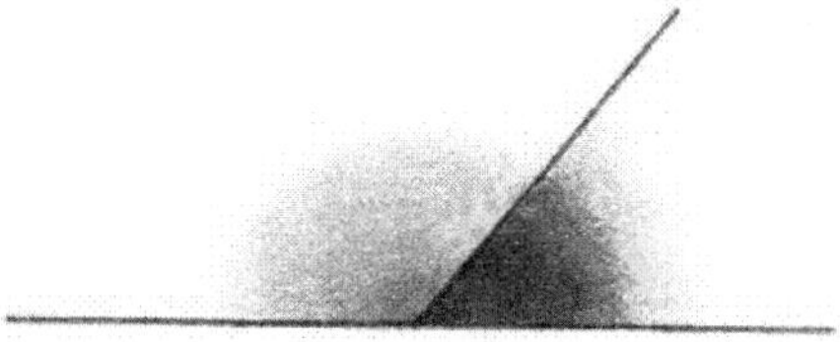

Vertical angles: A pair of non-adjacent angles formed by the intersection of two straight lines. Two vertical angles have the same measure.

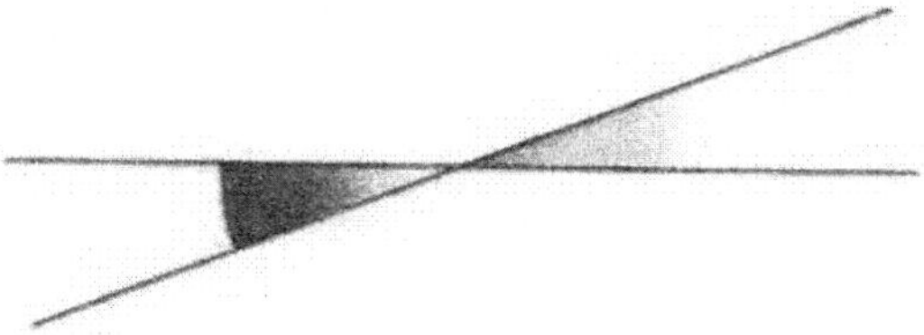

Vertical angles are **equal.**

Corresponding angles: Corresponding angles are created where a transversal line crosses a pair of parallel lines.

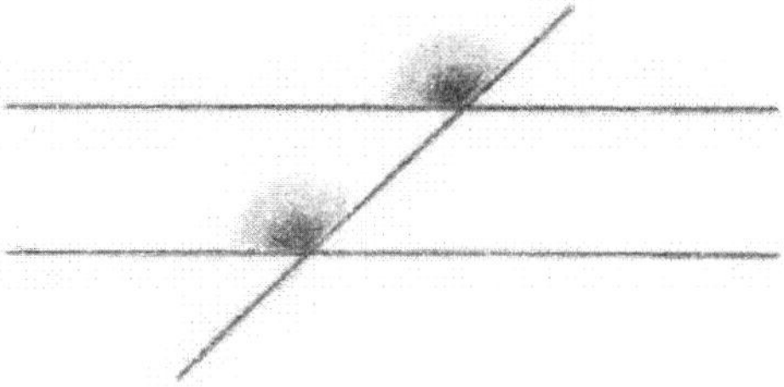

Corresponding angles are **equal.**

These angles are sometimes named **F** angles.

Alternate Angles: They can be classified into external and internal alternate angles.

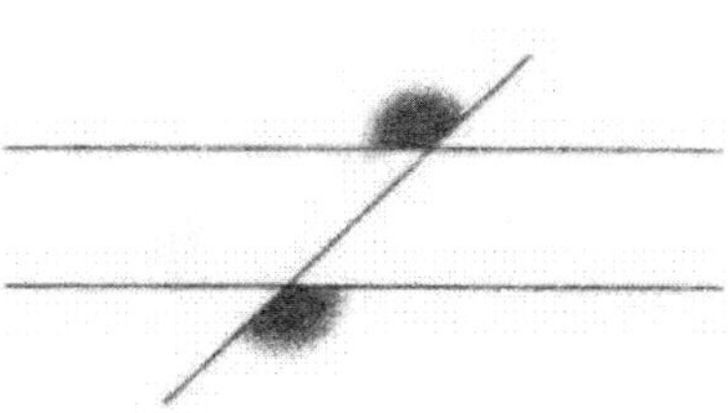

External alternate

Alternate angles are **equal.**

These angles are sometimes named **Z** angles.

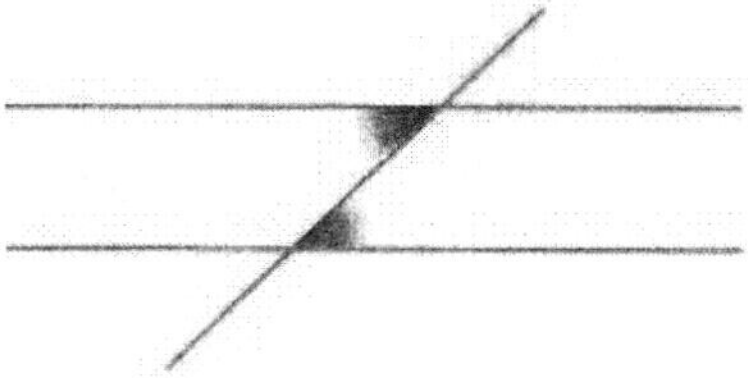

Internal alternate

Exercises

1. Find the missing angles in each diagram.

a) 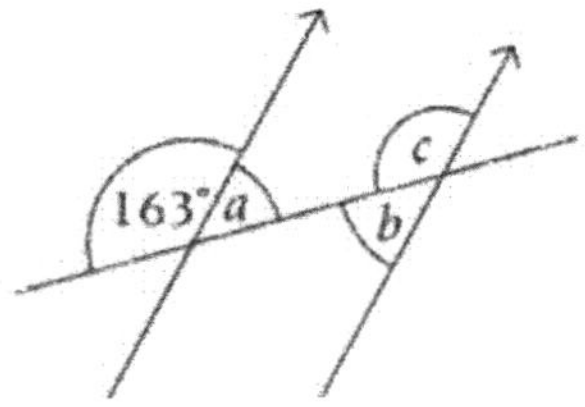

b) 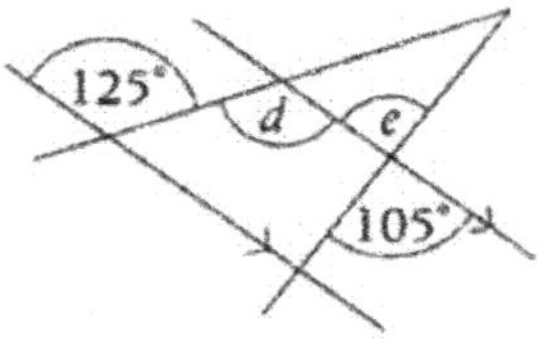

c) 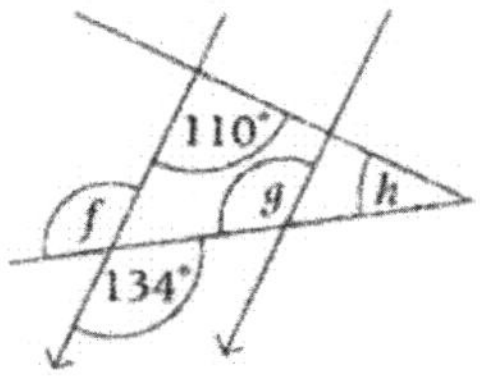

2. Angles in polygons

A **polygon** is a closed shape with three or more straight sides.

In a **regular polygon**, all its sides have the same length and all its interior angles are identical.

Polygons are named with prefixes indicating their number of sides: For example, figure shows a pentagon.

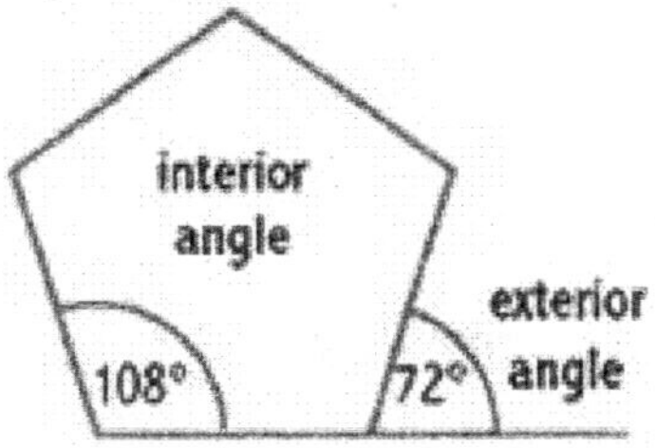

Interior angles are inside the polygon.

Exterior angles are made by extending each side in the same direction. Exterior angles are outside the polygon.

For a polygon with n sides the ***interior angle sum = (n − 2) × 180°***

Example: Check previous rule for a) a triangle and b) a square.

Solution:

a) A triangle has n = 3 sides, so, its interior angles sum is $(3-2)\cdot 180 = 180°$, as you already knew.

b) You know interior angles sum for a square is 360°. Let's check it: n = 4 sides, so, $(4-2)\cdot 180 = 360°$.

Exercises

2. Find the missing angles in these irregular polygons.

a)

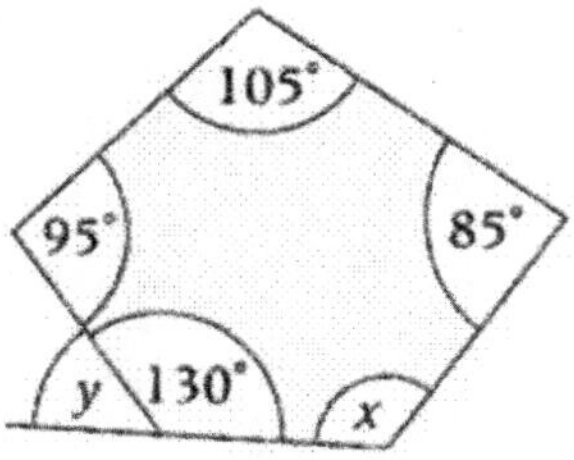

b)

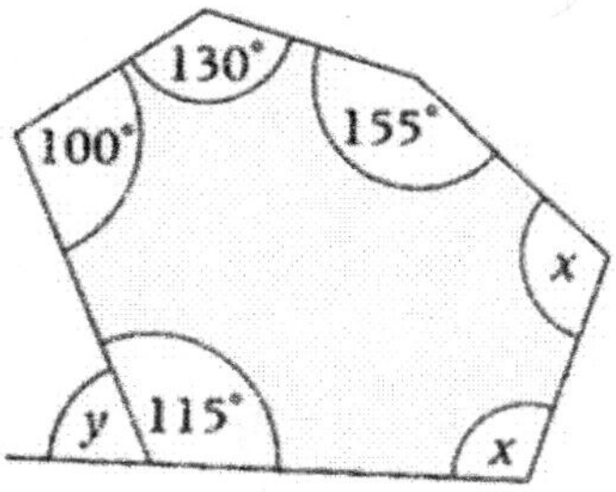

3. Calculate the unknown angles:

a)

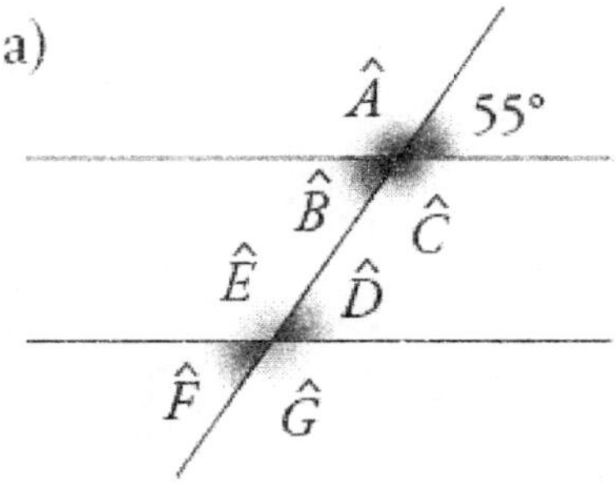

b)

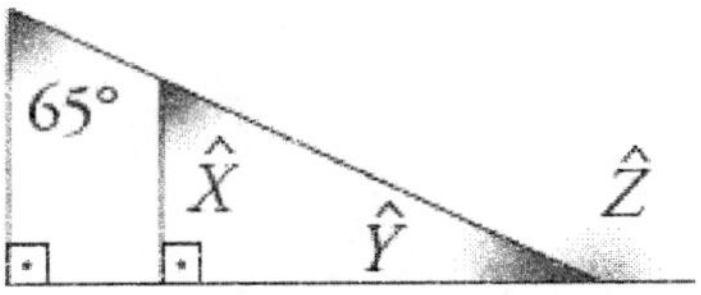

4. Calculate the unknown angles:

a)

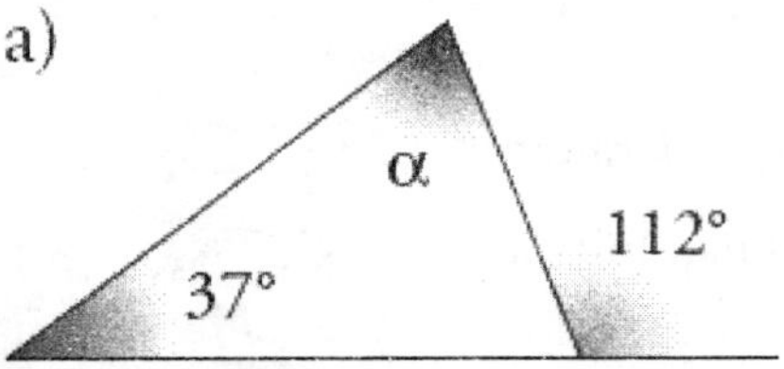

b)

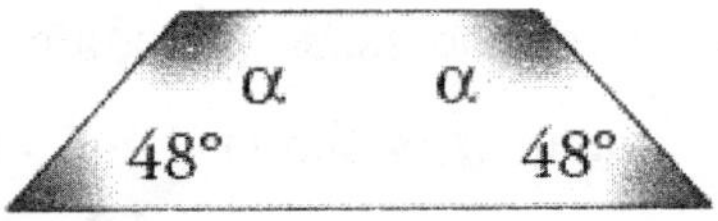

c)

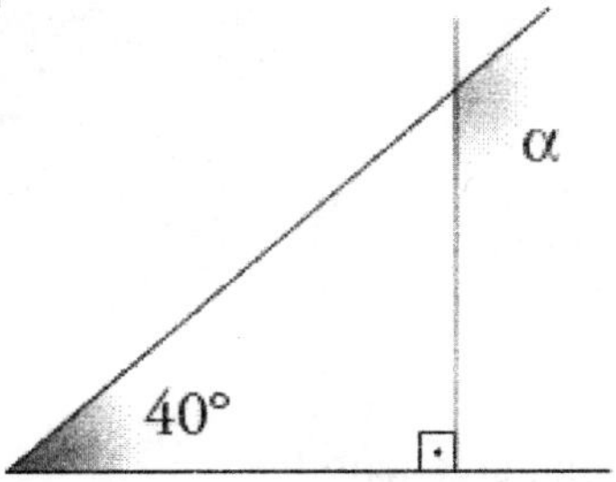

d)

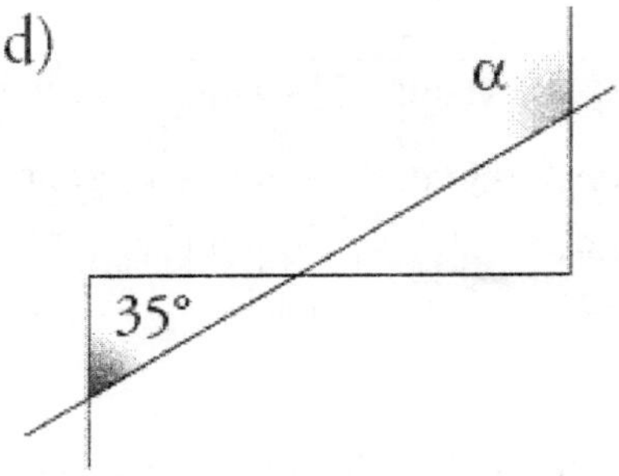

Exercises

5. Calculate the unknown angles:

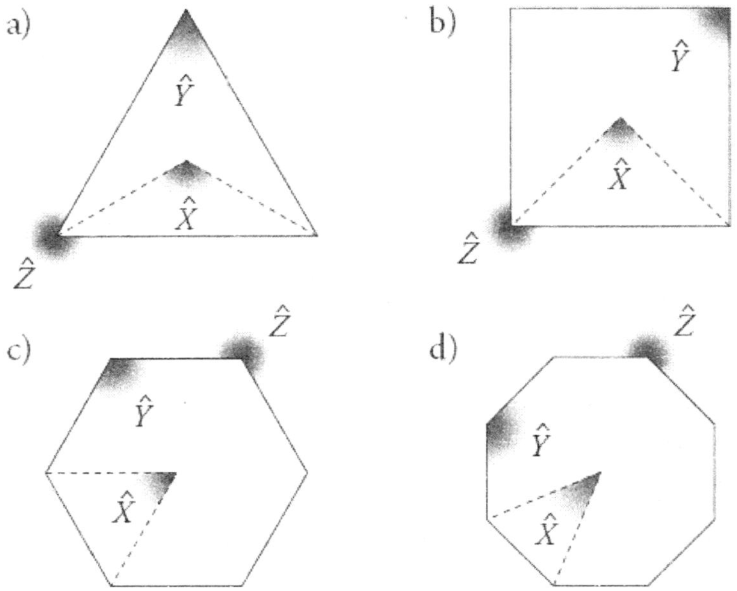

3. Similarity

Two figures that have the same shape are said to be **similar**. There are lots of examples of similar figures in quotidian life: an island and its map, a person and his photograph, etc.

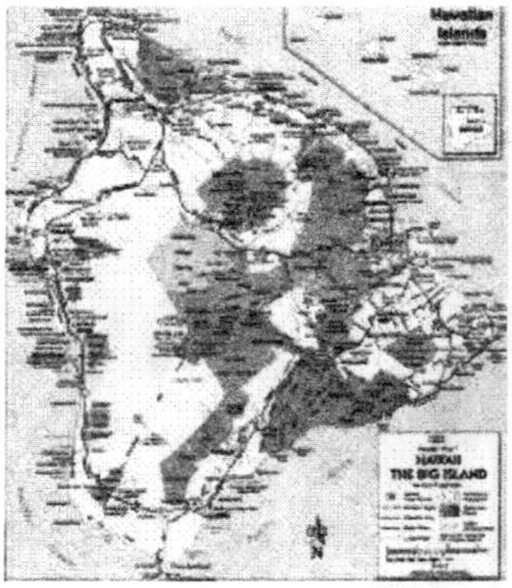

Two shapes are said to be ***similar shapes*** if:
- corresponding pairs of angles are equal
- corresponding pairs of sides are in the same ***ratio***.

You can use the ratio between similar shapes to find missing side lengths.

Example: Find unknown sides of following figures, knowing they are similar.

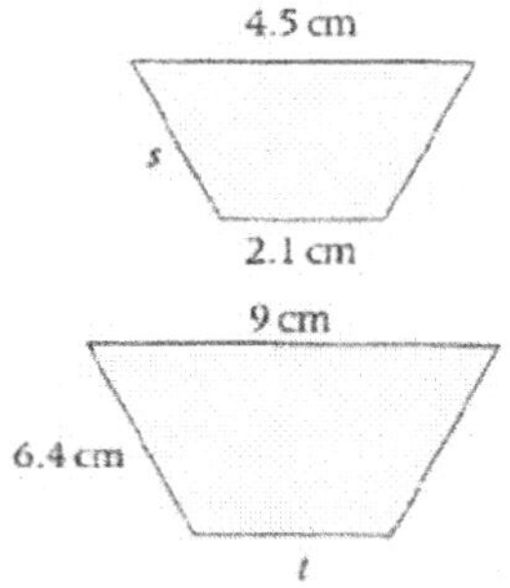

Solution:

The ratio larger/smaller is $\frac{9}{4.5} = 2$

Side s corresponds to side 6.4 cm $\rightarrow$ $\frac{6.4}{s} = 2 \Rightarrow s = \frac{6.4}{2} = 3.2\,cm$

Side t corresponds to side 2.1 cm $\rightarrow$ $\frac{t}{2.1} = 2 \Rightarrow t = 2.1 \cdot 2 = 4.2\,cm$

Exercises

6. These two parallelograms are similar. Find the perimeter of the smaller parallelogram.

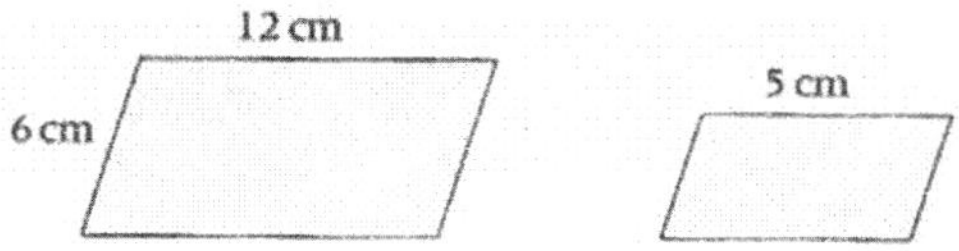

7. Measures of a photograph are 15 cm x 10 cm and it is surrounded by a frame being 2 cm wide. Are interior and exterior rectangles similar?

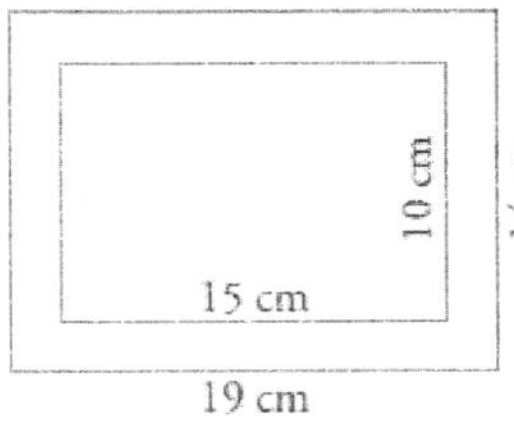

Scales. A ratio is used in scale drawings of maps and buildings.

Scale of a drawing = Drawing length : Actual length

A scale is usually expressed without mentioning units as in 1 : 100 000

A scale of 1 : 100 000 means that the real distance is 100 000 times the length of 1 unit on the map or drawing.

Exercises

8. A particular map shows a scale of 1 : 500. What is the actual distance if the map distance is 8 cm?

9. A particular map shows a scale of 1 cm : 5 km. What would the map distance (in cm) be if the actual distance is 14 km?

10. Actual distance of 900 km is represented on a map by 6 cm. What is its scale?

4. Similar triangles

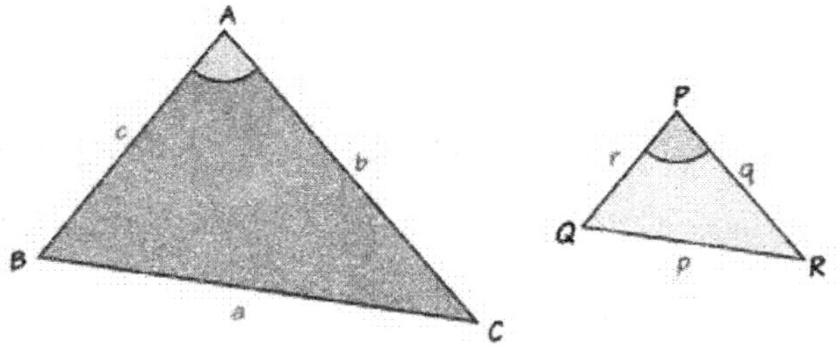

In similar triangles:

- corresponding pairs of angles are equal

A = P , B = Q , C = R

- corresponding pairs of sides are in the same ratio

$$\frac{a}{p} = \frac{c}{r} = \frac{b}{q}$$

If a line is drawn parallel to one side of a triangle, the smaller triangle and the larger one are *similar*. These two triangles are said to be in a ***Thales position***.

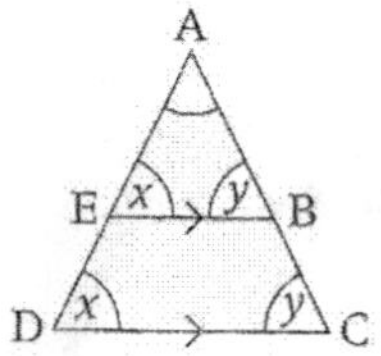

Example: Find the length CD in this triangle.

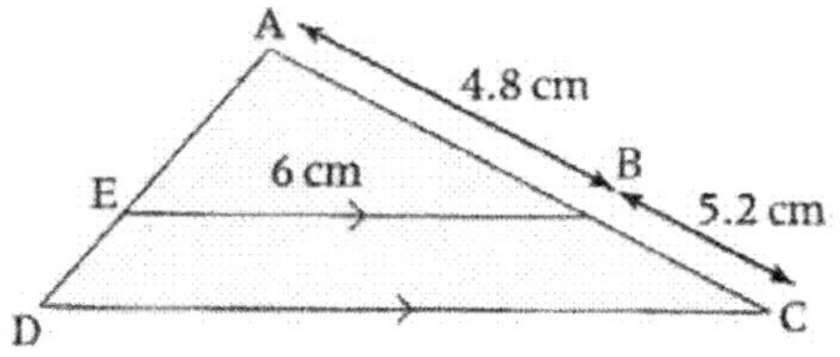

Solution:

Triangle ACD is similar to triangle ABE

$$\frac{CD}{6} = \frac{10}{4.8} \Rightarrow CD = \frac{6 \cdot 10}{4.8} = 12.5\ cm$$

Exercises

11. EB is parallel to DC. Work out the perimeter of: a) the triangle ABE and b) the trapezium EBCD.

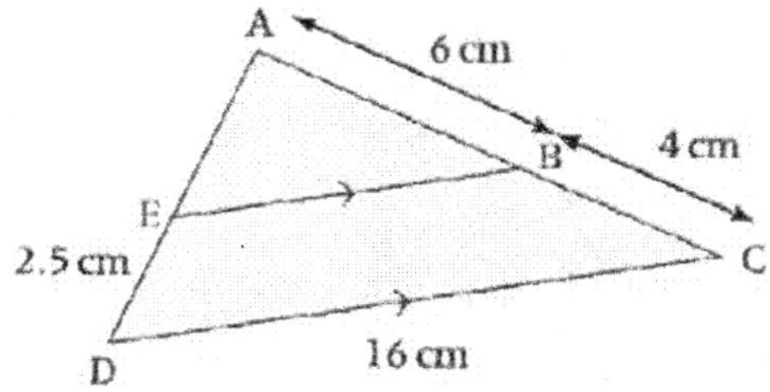

How to decide if two triangles are similar?

Two triangles are similar if they have all their corresponding pairs of angles equal and corresponding sides are in the same ratio. But we don't have to know all the sides and all the angles … two or three out of the six is enough.

There are three ways to find if two triangles are similar:

- If two triangles have two of their angles equal, the triangles are similar.
- If two triangles have two of their sides in the same ratio and the included angles are also equal, then the triangles are similar.
- If two triangles have three pairs of sides in the same ratio, then the triangles are similar.

Exercises

12. Check that the following pairs of triangles are similar.

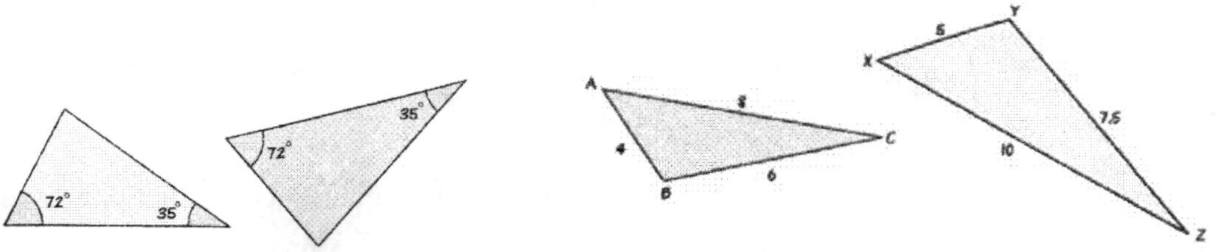

5. Pythagoras' Theorem

As you know, a **right triangle** or a *right angled triangle* is a triangle with one right (90°) angle.

The **hypotenuse** of a right triangle is the side that is opposite the right angle.

The **legs** of a right triangle are the sides adjacent to the right angle.

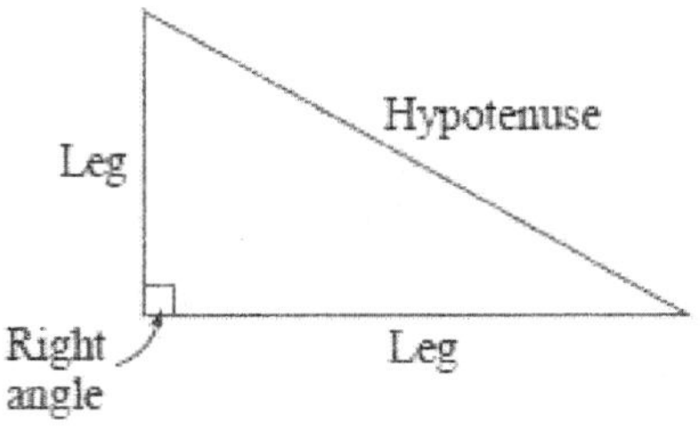

When drawing a right triangle, right angle is indicated by a little square (□), in order to avoid doubts about its measure.

Thousands of years ago, a Greek named **Pythagoras** discovered that the lengths of the sides of a right triangle have a special relationship. This relationship is known as the Pythagorean Theorem.

The ***Pythagorean Theorem*** enables us to calculate the length of the third side of a right triangle when we know the lengths of the other two sides.

Pythagorean Theorem

In a right triangle, the sum of the squares of the legs is equal to the square of the hypotenuse

$(\text{leg})^2 + (\text{leg})^2 = (\text{hypotenuse})^2$ or $a^2 + b^2 = c^2$

Exercises

13. In following right triangles, legs are given and we want to know the hypotenuse.

a) 37 cm and 45 cm b) 16 cm and 30 cm.

14. In following right triangles, one leg and hypotenuse are given and we want to know the other leg. a) 45 cm and 37 cm b) 39 cm and 15 cm.

15. In a rhombus, we know one diagonal, 24 cm, and its side, 13 cm. Find out the other diagonal.

16. A circumference has radius 15 cm. A straight line, r, intersects the circumference in two points, A and B. Distance between A and B is 18 cm. What is the distance from the circumference center to the straight line?

How to know if a triangle is a right one?

Pythagoras' Theorem can be used to determine if a triangle is acute, right or obtuse:

> If a and b are the shortest sides and c, the longest one:
> - If $a^2 + b^2 < c^2$, it is an acute triangle.
> - If $a^2 + b^2 = c^2$, it is a right triangle.
> - If $a^2 + b^2 > c^2$, it is an obtuse triangle.

Exercises

17. Classify into right angled, acute or obtuse, the triangles with sides:

a) 7 cm, 8 cm, 11 cm b) 11 cm, 17 cm, 15 cm

c) 34 m, 16 m, 30 m d) 65 m, 72 m, 97 m.

18. Find out the radius of this circumference (Fig. 1), knowing: OP = 39 cm, PT = 36 cm.

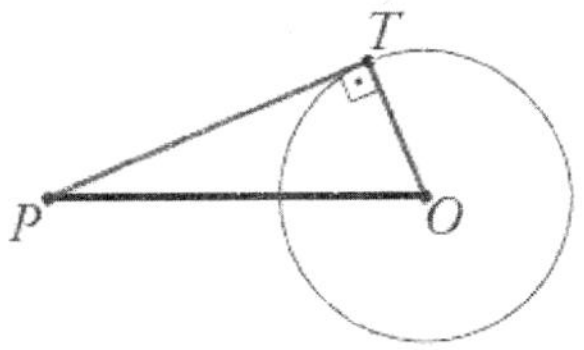

Figure 1

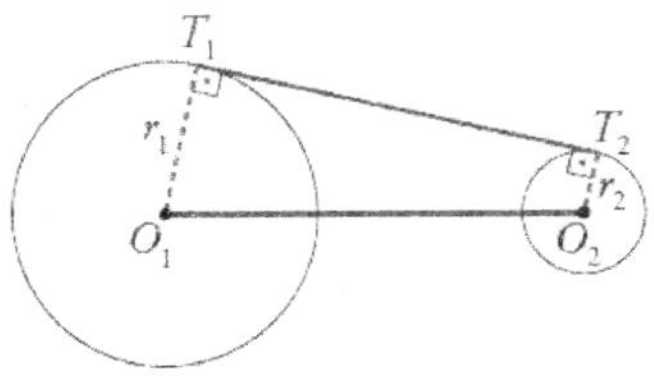

Figure 2

19. In Figure 2, $r_1 = 15$ cm, $r_2 = 6$ cm, $O_1O_2 = 41$ cm. Calculate the length of T_1T_2.

20. Check if triangle with sides 29 cm, 35 cm and 48 cm is right, acute or obtuse. Calculate the length of its height on the largest side.

21. Sides of a trapezium measure 13 m, 20 m, 19 m and 40 m, respectively. Two last sides are parallel. Calculate the height of the trapezium.

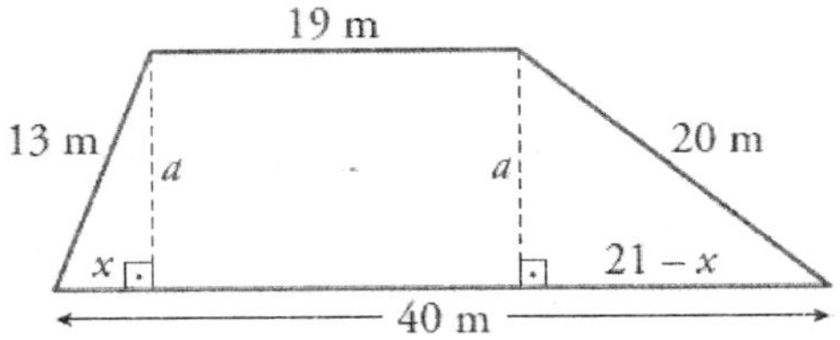

6. Areas of polygons

Area of a polygon is quantity of surface existing into the closed zone delimited by sides. Some of the expressions used to calculate areas of polygons might be known by you.

They are shown in the next figure. Anyway, we will calculate areas of polygons (regular or irregular) by cutting them into three basic shapes: rectangles, triangles and circles (or portions of them, as semicircles, etc).

KNOWN AREAS

RECTANGLE

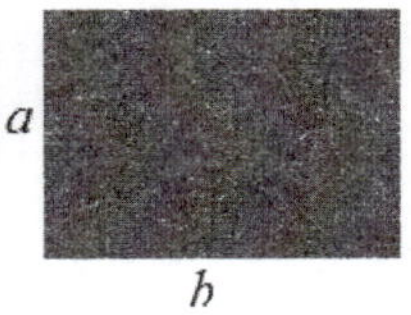

$$A = b \cdot a$$

SQUARE

$$A = l^2$$

PARALLELOGRAM

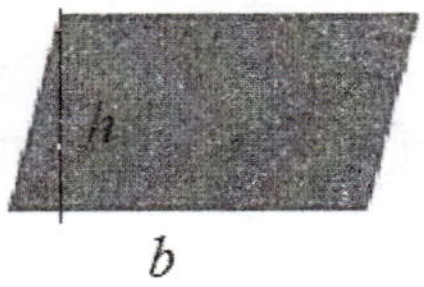

$$A = b \cdot a$$

RHOMBUS

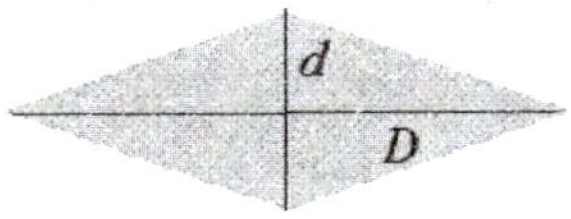

$$A = \frac{D \cdot d}{2}$$

TRAPEZIUM

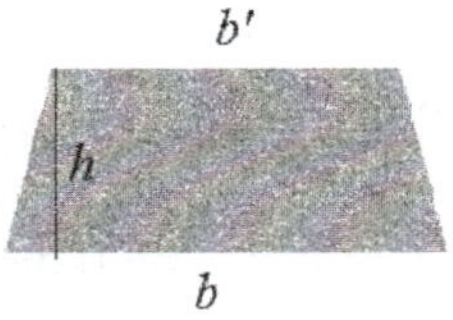

$$A = \frac{b + b'}{2} \cdot h$$

REGULAR POLYGON

$$A = \frac{Per \cdot ap}{2}$$

TRIANGLE

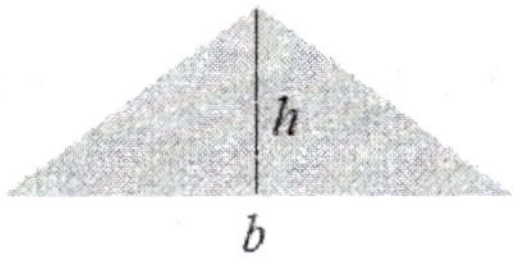

$$A = \frac{b \cdot h}{2}$$

RIGHT TRIANGLE

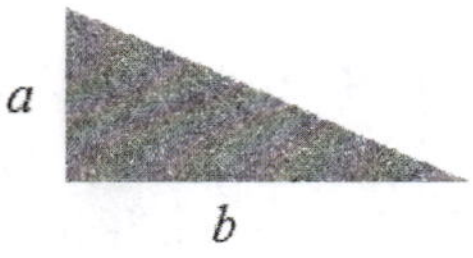

$$A = \frac{b \cdot a}{2}$$

Heron´s formulae

There is a useful expression used to calculate area of triangles (regular or irregular), knowing their three sides. It was developed by Heron, so, it is known by his name.

Heron´s formulae

If a triangle has sides *a, b* and *c*,

- its perimeter, *p*, is $p = a + b + c$,
- its semi-perimeter, *s*, is, $s = \frac{p}{2}$

And its area can be calculated as $\boxed{A = \sqrt{s \cdot (s-a) \cdot (s-b) \cdot (s-c)}}$ *(HERON'S FORMULAE)*

Exercises

22. Calculate the area of a triangle whose sides measure 10 m, 17 m and 21 m.

23. Calculate the area of the regular hexagon with side 10 cm.

24. Calculate the area of a rhombus with side 3 dm, knowing its diagonal, 46 cm.

25. Two sides of an isosceles triangle measure 30 cm and 13 cm. Calculate its area.

7. Areas of curved figures

KNOWN AREAS

CIRCLE | CIRCLE SECTOR | ANNULUS (Circle crown) | ELLIPSE

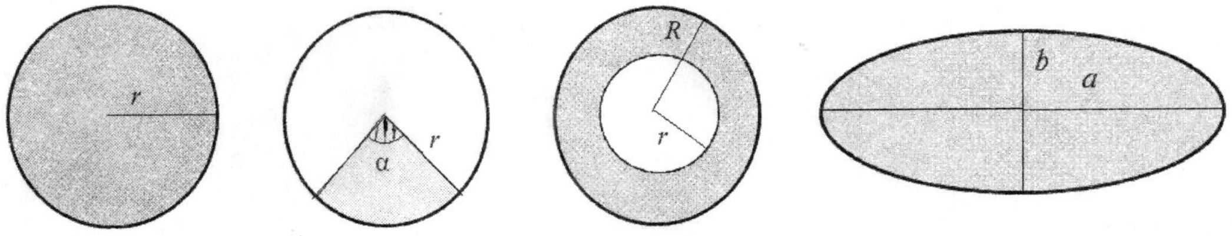

$A = \pi \cdot r^2$ | $A = \pi \cdot r^2 \cdot \frac{\alpha}{360}$ | $A = \pi \cdot (R^2 - r^2)$ | $A = \pi \cdot a \cdot b$

Exercises

26. Calculate the area of the coloured zone of the following figures:

a) b) c) d)

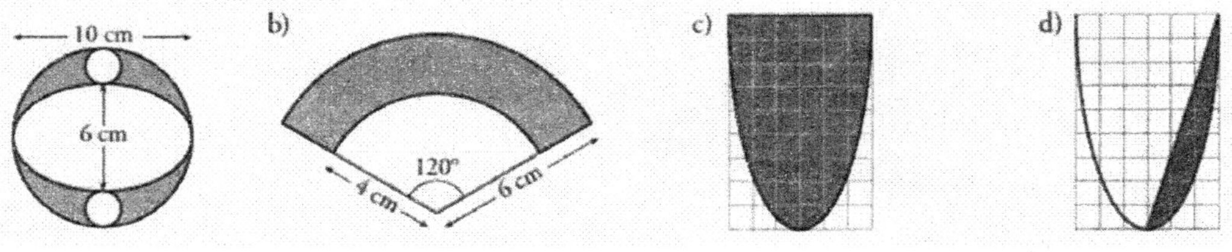

Review exercises

Similar shapes

1. On a map whose scale is 1:1 500 000, distance between two towns is 3,5 cm.

a) What is the actual distance between them?

b) What is the distance, on that map, between two towns at 250 km?

2. Calculate actual distance that is represented by 13 cm on a map with scale 1: 1800000.

Similarities in triangles

3. Calculate the lengths of the sides *a* and *b* knowing these triangles have their sides parallel:

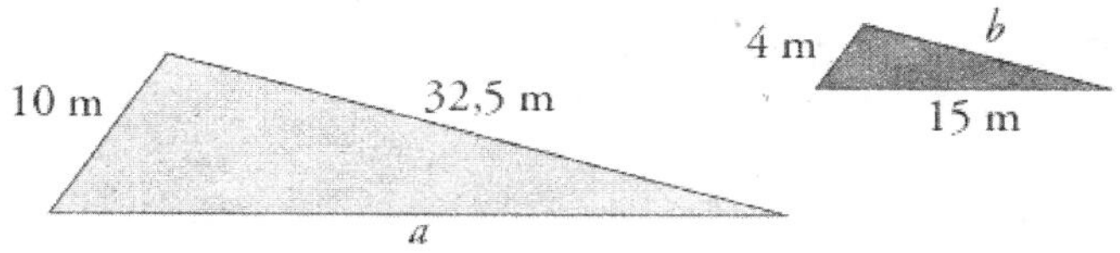

4. Thales was able to determine the height of a pyramid only by measuring its shadow, 18 m, and the shadow projected by his stick, 0.5 m, at the same time. What was its height if his stick was 1.70 m?

Pythagoras' theorem

5. Find out the value of x in the following shapes:

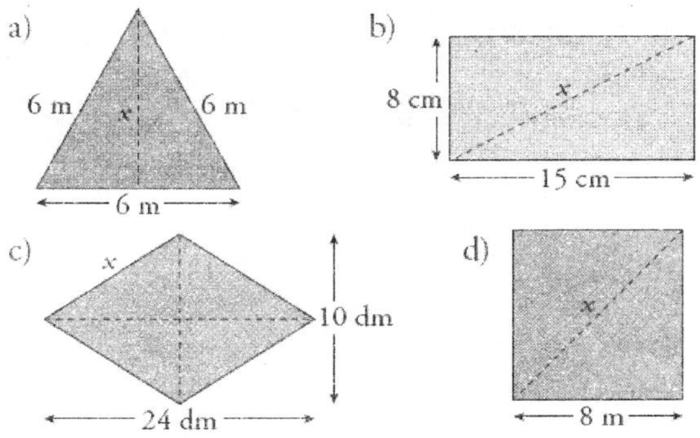

6. Classify into right angled, acute or obtuse, the triangles with sides:

a) 11 m, 13 m, 20 m. b) 20 m, 21 m, 29 m. c) 25 m, 29 m, 36 m.

d) 7 m, 24 m, 25 m.

7. Work out the value of x in each shape:

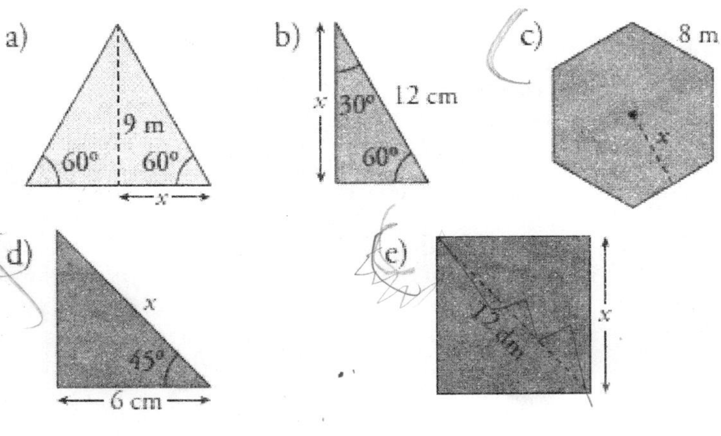

8. Calculate the diagonal of a square with side 6 m.

9. Calculate the diagonal of a rectangle with sides 3m and 6 m.

10. Calculate the side of a square, with a 10 cm diagonal.

11. Calculate the side of a rhombus with diagonals 5 m and 8 m.

12. Calculate the height of an equilateral rectangle with side 5 cm.

13. Calculate the side of an equilateral rectangle knowing its height is 6 m.

14. In a right angled triangle, legs measure 9 cm and 12 cm. In another right angled triangle, a leg measures 14.4 cm, and hypotenuse, 15 cm. Which of them has a higher perimeter?

15. Diagonal of a rectangle with sides 7 cm and 24 cm has the same dimensions that the side of a square. Find out the diagonal of this square.

16. Diagonal of a rectangle measures 37 cm, and one of its sides, 12 cm. Calculate its perimeter and its area.

Polygon areas

17. Calculate the area of the following figures:

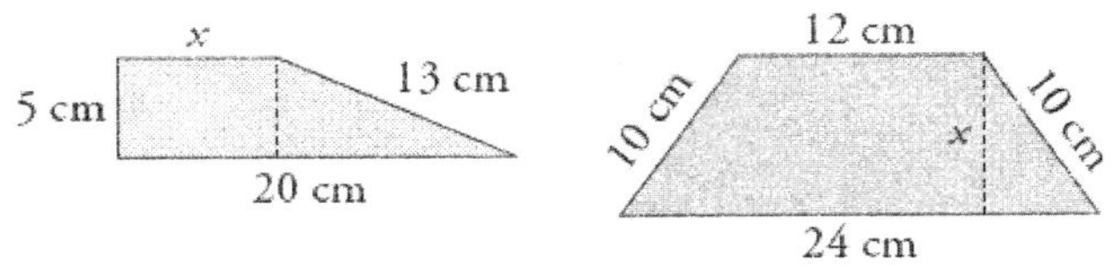

18. Calculate the area of a regular octagon having a perimeter of 64 cm and an apothem of 4 cm.

19. Calculate the area of coloured zones:

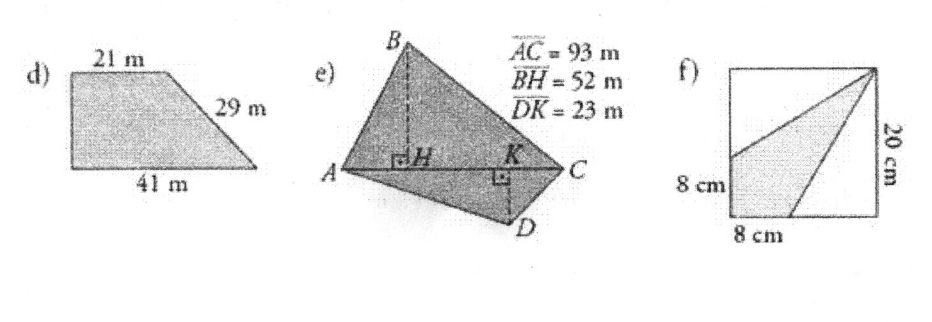

Curved shapes

20. Calculate the area of a circle segment with angle = 90° in a circle with radius = 18 cm.

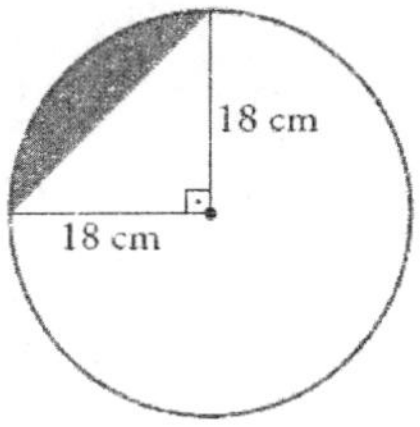

21. Calculate the area of the following coloured shapes:

a)

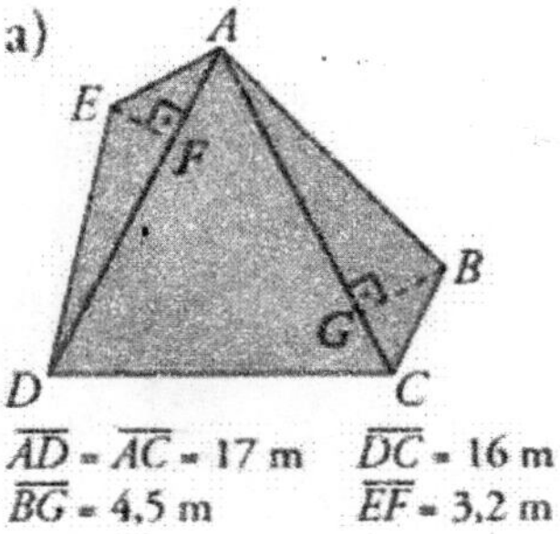

b)

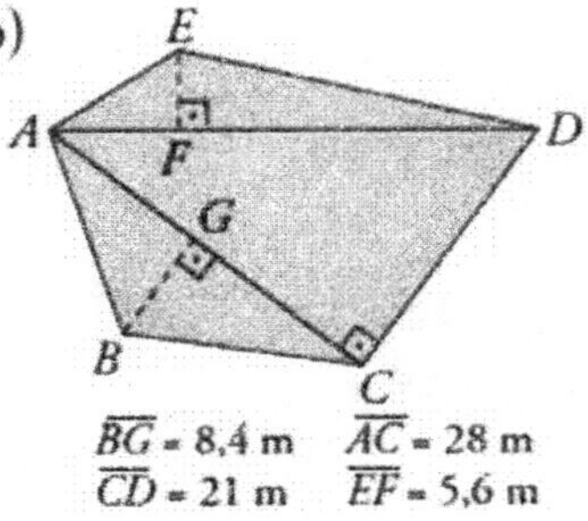

c)

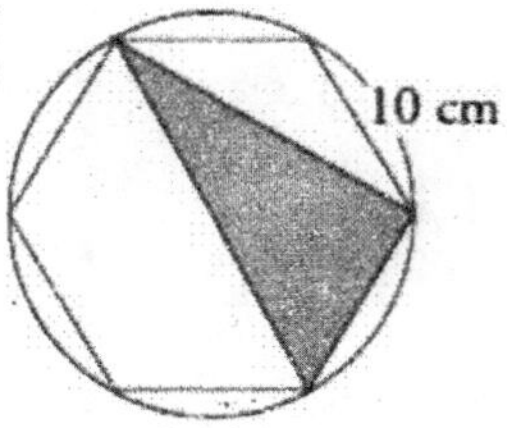

d)

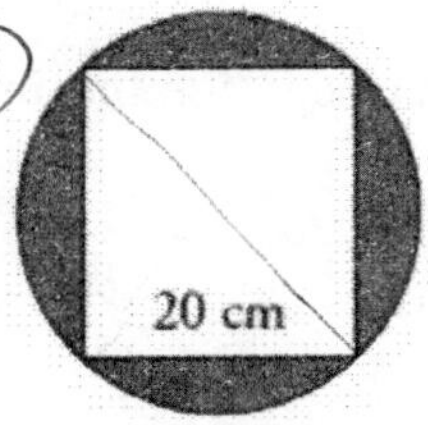

e)

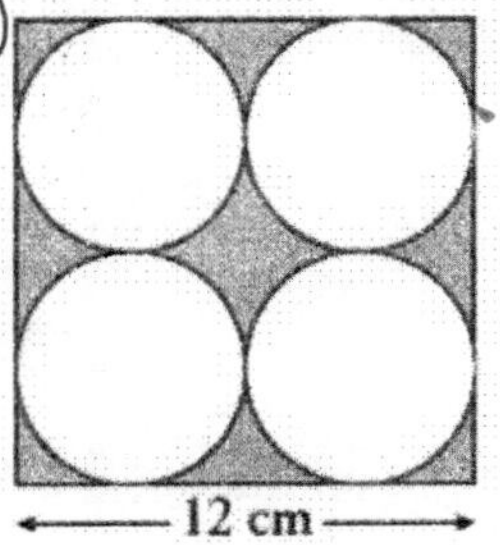

f)

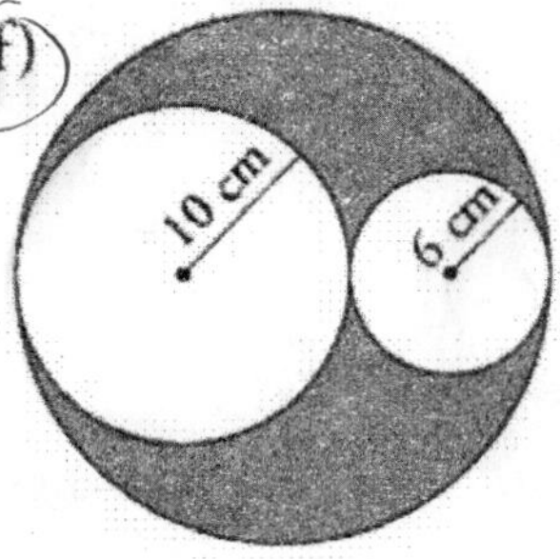

22. A circle has a circumference of 56.52 cm. Draw its circumscribed and inscribed squares, and calculate their perimeters and areas. (Consider $\pi = 3{,}14$).

23. Find out the areas of circle sectors with radius of 15 cm and angles:

a) 90° b) 120° c) 65° d) 140°

24. Calculate the areas of coloured zones in each of these squares, having sides of 12 cm:

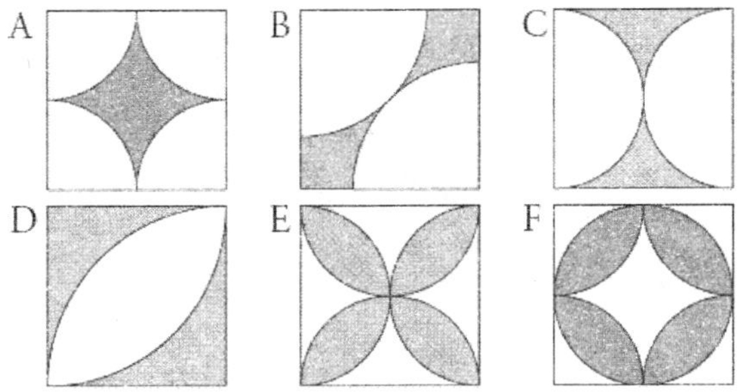

25. Calculate the area of the coloured zones:

a)

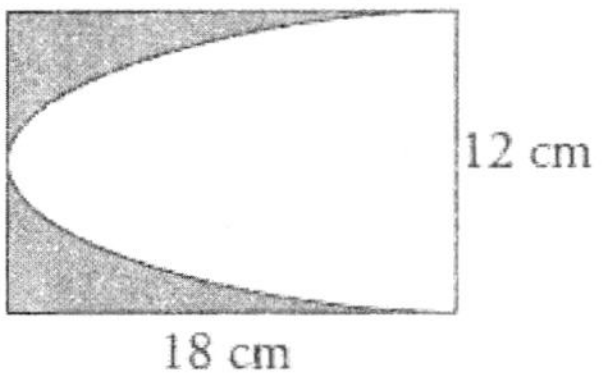

b)

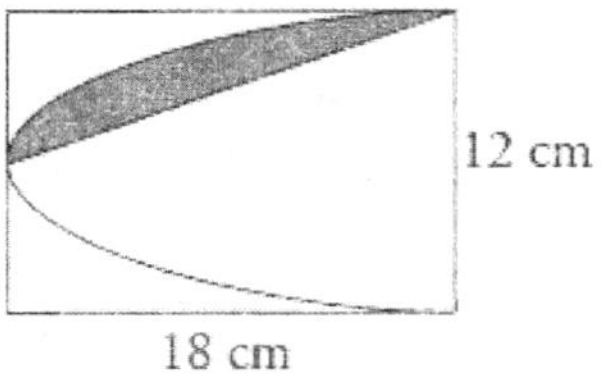

26. Diagonals of inscribed rhombus in the ellipse measure 16 cm and 30 cm. Calculate the area of the coloured zone.

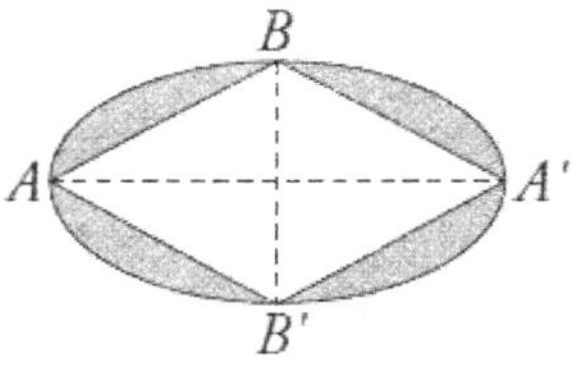

Revision

27. In a triangle *ABC*, base *AB* measures 5.7 m and height on that base is 9.5 m. What is the area of another triangle, A′B′C′, similar to this one, in which *A'B′* = 4.14 m?

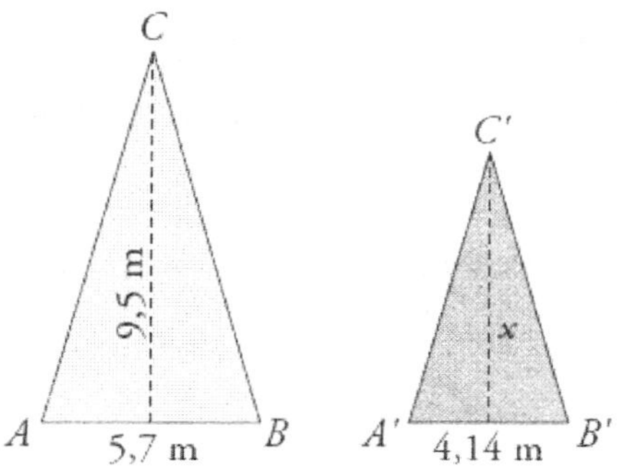

28. If BD side is parallel to AE, and $AC = 15$ cm, $CE = 11$ cm, $BD = 6.4$ cm, $AE = 18$ cm:

a) Calculate CD and BC.

b) If $A = 37°$ and $C = 80°$, calculate E, B and D.

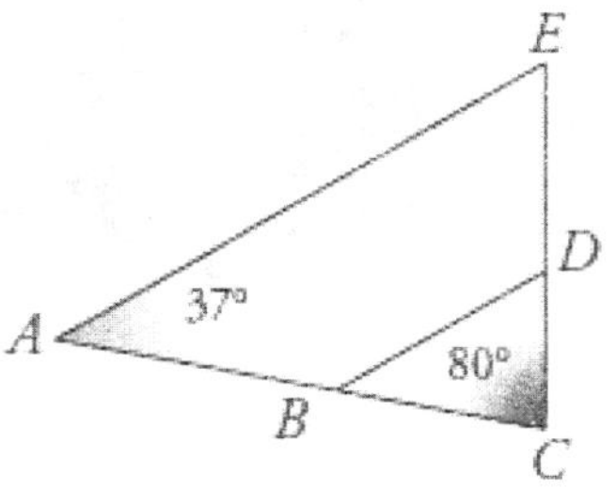

29. a) Why are triangles APQ y ACB *similar*?

b) Calculate BQ.

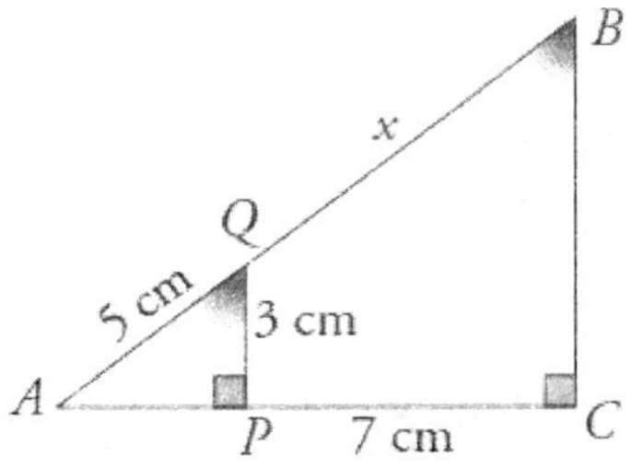

30. If $DF = 4$ cm, what is the area and perimeter of trapezium $EFAC$?

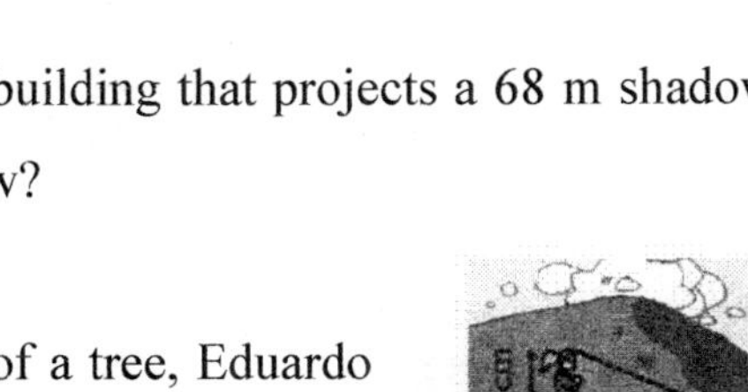

31. What is the height of a building that projects a 68 m shadow at the same time that a 1.65 m person projects a 2 m shadow?

32. To calculate the height of a tree, Eduardo (1.62 m) looks at the reflected top of the tree, as shown in the picture. What is the height of the tree?

33. What is the depth of a well if its width is 1.5 m and, if moving 0.5 m farer, from a height of 1.7 m, you see the edge of the well on the same straight line than the ground?

34. Calculate the height of the following triangle, by applying Pythagoras' theorem to both right triangles. Calculate its area.

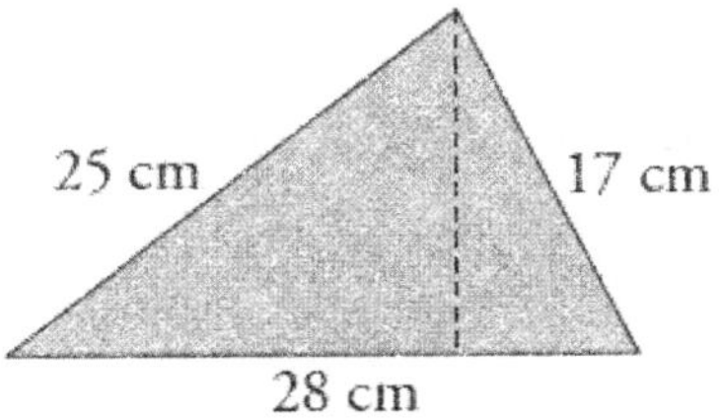

35. Area of a circle crown is 20π cm^2, and inner circumference measures 8π cm. Calculate radius of outer circumference.

36. Calculate the height and area of the following trapezium.

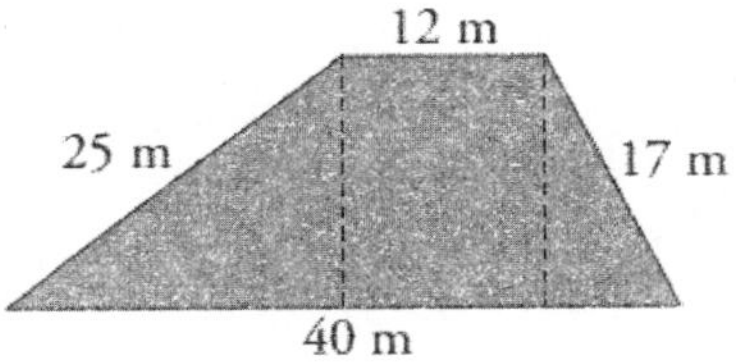

37. a) Calculate the radius of the following circumference.

b) What will be the length of a chord whose distance to the center is 2.9 cm?

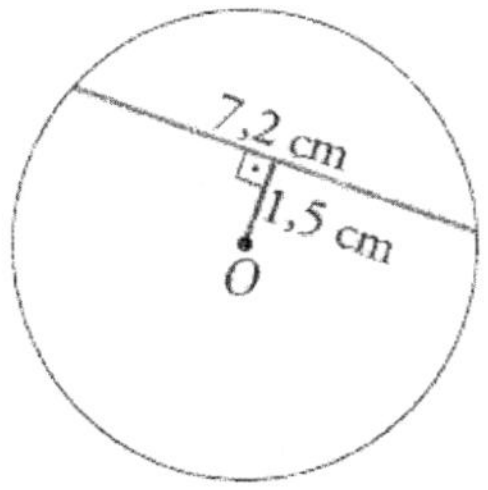

38. Calculate:

a) The length of *PT*.

b) The area of the coloured zone.

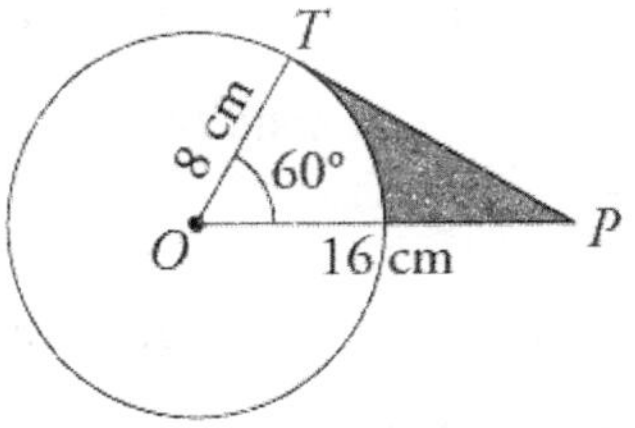

39. Calculate the area of the curved triangle existing between three tangent circumferences with radius 5 cm.

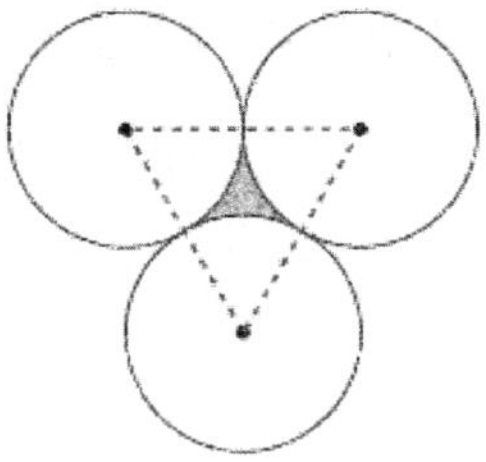

Unit 10.- Movements in the plane

1. Vectors

Magnitudes like mass, temperature, time, are perfectly defined by giving a number and its unit. For example, we say mass of an apple is 125 g, or we say our class takes 55 minutes. We do not need to say anything more.

But, imagine you are near a car moving with speed of 70 km/h. That magnitude is not totally defined, because if car is going away, you can stay where you are, but if it is going towards your position, you should move quickly from there. So, we need some more information to define the magnitude *speed*. Or, for example, imagine you apply a force of 10 N on the head of someone in a swimming pool. Consequences are going to be very different if you push him down or if you pull him up, do you agree?

So, we are seeing there are two types of magnitudes:

Scalar magnitudes are those that are perfectly defined only by giving their magnitude and unit. For example, *mass, time, temperature, density,…*

Vectorial magnitudes are those that apart from magnitude and unit, need some more information to be defined. They need information like *where* they are applied, *direction* and *sense* of application,.... For example, *speed* and *force* are vectorial magnitudes.

Vectorial magnitudes are usually associated to a ***vector***. A vector is like an *arrow*, but, mathematically, it is defined as a mathematical object, which is determined by the following elements:

Elements of a vector:
- ***Application point:*** where it is applied.
- ***Magnitude:*** its length (the higher the applied force is, the longer the vector is).
- ***Direction:*** straight line in which it is.
- **Sense:** a straight line has two senses. Sense is determined by the head of the arrow.

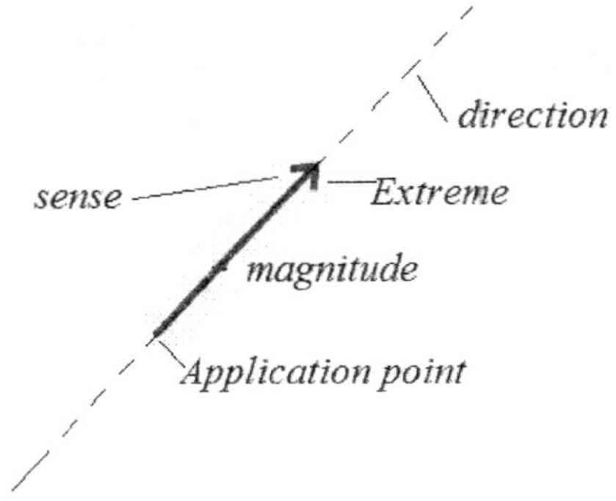

Notice that in quotidian life, we use the word *"direction"* in a different way. We say: "I am going in the direction to Madrid", meaning the *sense*.

In mathematical terms, *direction* would be the straight road joining Murcia and Madrid. And in that road, there are cars going in the *sense* Murcia → Madrid, and other ones in the *sense* Madrid → Murcia.

Coordinates and magnitude of a vector

A vector, $\boldsymbol{v}$, is numerically determined by its coordinates, a pair of numbers, (v_x, v_y).

If vector $\boldsymbol{v}$, is determined by points A and B, with coordinates $A(a_x, a_y)$ and $B(b_x, b_y)$, its coordinates can be determined as $\boldsymbol{v} = \boldsymbol{AB} = (b_x - a_x, b_y - a_y)$

Example: Calculate the coordinates of vector **AB**, with A(2, 7) and B(6, 12), and draw it.

Solution: AB = (6 – 2, 12 – 7) = $\boxed{(4, 5)}$

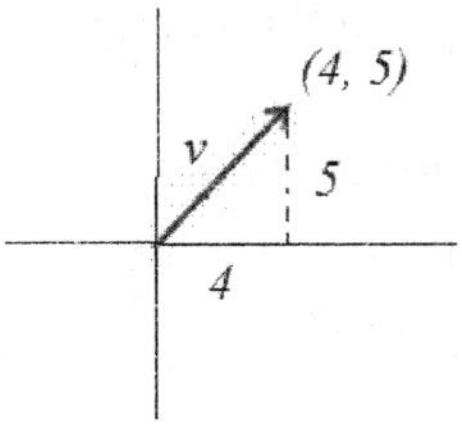

Example: Coordinates of vector **AB** are (2, 1). Knowing coordinates of point A(5, 2), calculate coordinates of point B.

Solution:

$$\left.\begin{array}{l} A=(5,2) \\ B=(b_x,b_y) \end{array}\right\} \rightarrow AB=(2,1)=(b_x-5,\ b_y-2) \rightarrow \left\{\begin{array}{l} b_x-5=2 \rightarrow b_x=5+2=7 \\ b_y-2=1 \rightarrow b_y=2+1=3 \end{array}\right. \rightarrow \quad So,\ \ B(7,3)$$

Once we already know coordinates of a vector, we can know how long it. This is, its ***magnitude***. We are going to do it by using Pythagoras' theorem. As you can see in the figure below, coordinates v_x and v_y are the legs of a right triangle, whose hypotenuse is *magnitude* of the vector. So,

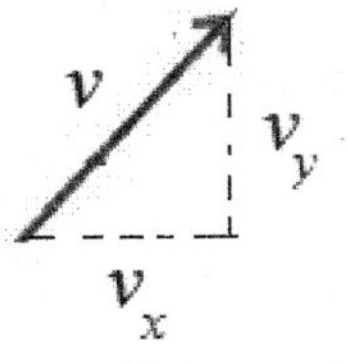

$$v^2 = v_x^{\ 2} + v_y^{\ 2} \quad \rightarrow$$

$$v = \sqrt{v_x^{\ 2} + v_y^{\ 2}}$$

Magnitude of a vector is calculated as $v = \sqrt{v_x^{\ 2} + v_y^{\ 2}}$

Example: Points C(2, 5) and D(0, 4) determine vector CD. Calculate a) its coordinates and b) its magnitude.

Solution:

a) **CD** = (0 – 2, 4 – 5) = $\boxed{(-2, -1)}$

b) $v = \sqrt{(-2)^2 + (-1)^2} = \sqrt{4+1} = \boxed{\sqrt{5}}$

Exercises	
	1. Vector **AB** is applied on point A(1,2) and its head is on B(5, 8). a) Draw both points and the vector. b) Calculate the coordinates of the vector. c) Determine its magnitude. **2.** Work out the radius of the circle with the center (8, -2) passing through the point (1, 4). **3.** Calculate the magnitude (or length) of these vectors: **AB** = (35, -21) and **CD** = (-10, 6) **4.** Given point A=(1, -2). Decide what is the nearest point from A, B or C, with B(4, 0) and C(0, 1). **5.** Work out the coordinates of D, so that the quadrilateral of vertex: A(-1, -2), B(4, -1), C(5, 2) and D; is a parallelogram. **6.** Given the vector **a** = (5, 3) and the point A = (4, -1), find the coordinates of the point B so that the vector **AB** is the same as the vector a.

Midpoint of a segment

Example: Segment AB is determined by points A(2, 5) and B(6, 1). Calculate the coordinates of the midpoint of this segment.

Solution:

Midpoint can be named M, with unknown coordinates M(m_x, m_y). Because of being the midpoint, two vectors **AM** and **MB** are equal. So,

AM = (m_x – 2, m_y – 5)

MB = (6 - m_x, 1 - m_y)

As **AM** = **MB**, we have:

$$(m_x - 2,\ m_y - 5) = (6 - m_x,\ 1 - m_y) \rightarrow \begin{array}{l} m_x - 2 = 6 - m_x \rightarrow 2 \cdot m_x = 8 \rightarrow m_x = 4 \\ m_y - 5 = 1 - m_y \rightarrow 2 \cdot m_y = 6 \rightarrow m_y = 3 \end{array}$$

So, midpoint of the segment is M(4, 3)

A different method: We can use a different equality to calculate midpoint of a segment. Notice that vector

AB is twice **AM**, so, **AB** = 2 · **AM**

AB = (4, – 4)

AM = (m_x – 2, m_y – 5)

As 2 · **AM** = **AB**, we have:

2 · (m_x – 2, m_y – 5) = (4, - 4) → $2 \cdot m_x - 4 = 4 \rightarrow 2 \cdot m_x = 8 \rightarrow m_x = 4$; $2 \cdot m_y - 10 = -4 \rightarrow 2 \cdot m_y = 6 \rightarrow m_y = 3$

So, midpoint of the segment is **M(4, 3)**

Exercises

7. Calculate central point of a segment determined by points (1,4) and (5,-8).

8. A town A is located at point (-2, 5) and another one, B, at point (4,10). They decide to build a shopping center in the central point of the road that joins them. Where will they build it?

9. Central point of a segment AB is located at point M(3,5). If A is (-4,2), where is B located?

10. Calculate the symmetric point of A(1,2) about B(-2,7). A scheme might be useful for you.

11. Given the segment determined by A(0,4) and B(9,-2), determine the location of two points that divide it into three similar portions.

12. A frog moves by jumping on a straight line. Its first jump starts at point A(-4, -4) and finihes at point B(2,1), where it starts a new jump. Determine the point in which it will finish its fifth jump from A.

Operations with vectors

Vectors can be added, subtracted, multiplied by a number, or a combination of these operations. Rules for these operations are shown in the following chart:

Operations with vectors Consider vectors $v(v_x, v_y)$ and $w(w_x, w_y)$,
- **Addition:** $v + w = (v_x+w_x, v_y+w_y)$.   - **Subtraction:** $v - w = (v_x-w_x, v_y-w_y)$.   - **Product by a number:** $a\cdot v = (a\cdot v_x, a\cdot v_y)$.

Exercises

13. Given vectors **u**(3,5) and **v**(-2,0), calculate: a) **u+v**; b) **u-v**; c) **3·v-u**

14. Given vectors **a** = (9, 3) and **b** = (-5, 4), calculate the coordinates of 2**a** + 3**b**.

15. Complete:

a) (……, 5) + (2, …..) = (-3,0)

b) 2·(0,3) + 3(…..,….) = (-3, 0)

c) (-4, …..) – (……., 2) = (-4,5)

d) (2,3) – 2·(….,….) = (-2, -7)

2. Geometric transformations

A ***Geometric transformation*** (or simply transformation) is a general term for four specific ways to manipulate a point, a line, or shape. The original shape of the object is called the ***pre-image*** and the final shape and position of the object is the ***image under the transformation***.

Lots of artists, mainly in Arabic art, use geometric transformations (named *mosaic* and *tessellations*) to create beautiful artistic objects. Alhambra, in Granada (Spain) is absolutely full of mosaic and tessellations.

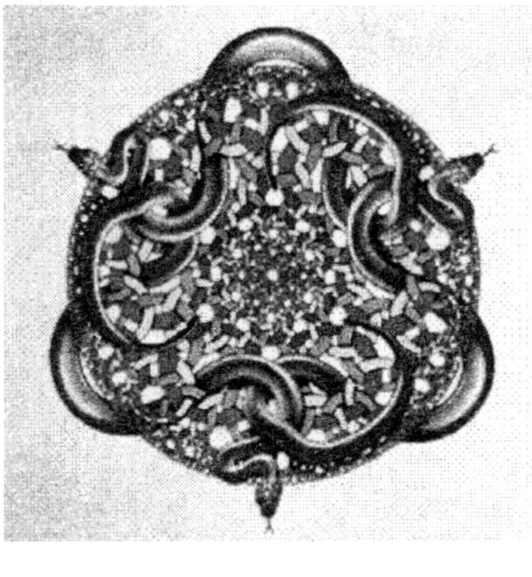

Types of transformations in math

- **Isometry**: It is a <u>*congruent*</u> transformation. This means that image and pre-image have the same side lengths and angle measurements. The following transformations maintain their mathematical congruence.
    - · Traslation (direct)
    - · Reflection (inverse)
        - · Point reflection
        - · Line reflection
    - · Rotation (direct)
- **Dilation:** on the other hand, a dilation is not an isometry because its image does not conserve length of sides.

Geometric transformations can be classified into *direct* and *indirect* transformations. ***Direct*** transformations are those in which orientation of pre-image is conserved.

3. Traslations

A ***traslation*** is considered a direct isometry because it not only maintains congruence, but it also, unlike reflections and rotations, preserves its orientation.

> A ***traslation*** is a transformation in which a graph or geometric figure is picked up and moved to another location without any change in size or orientation.
>
> The notation for traslation is $T(+a,+b)$ where a and b represent how much you slide in the x and the y directions, respectively.

Following figures show examples of geometric traslations.

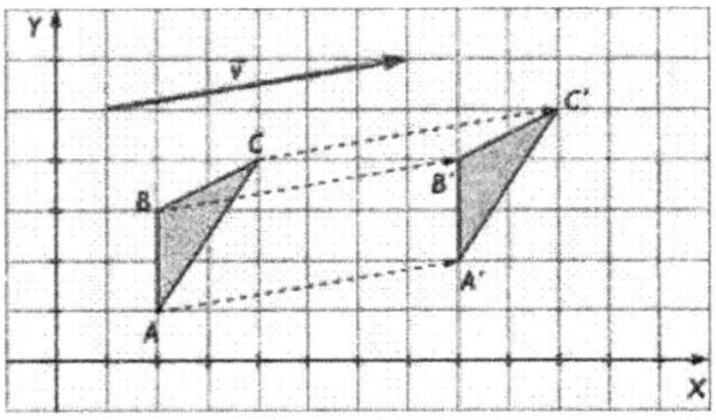

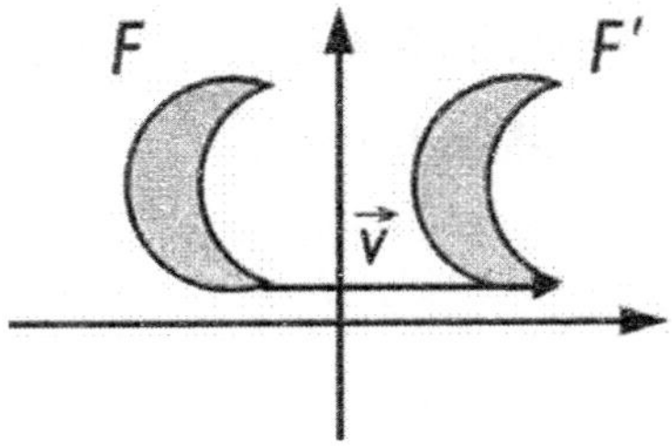

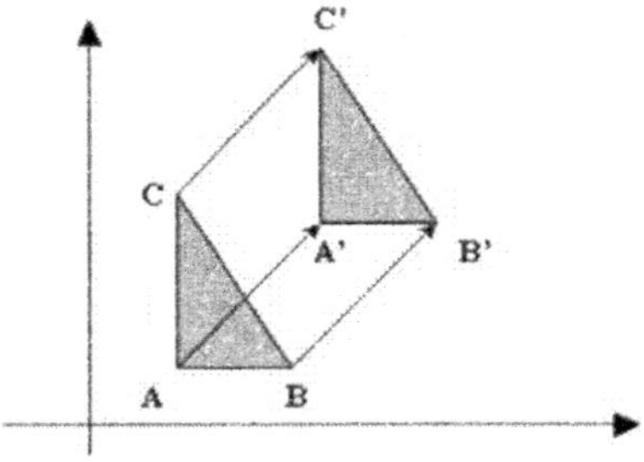

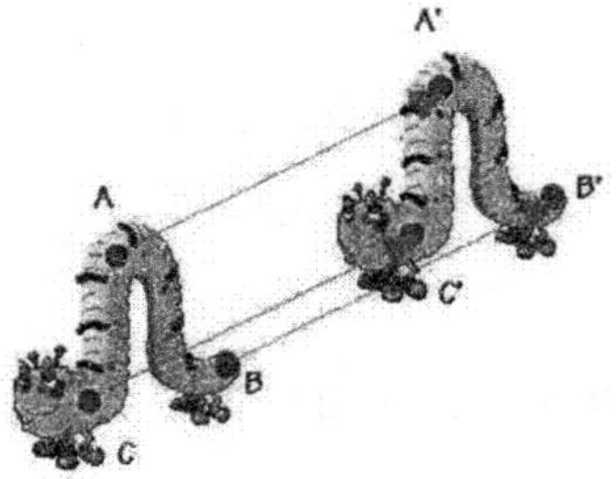

Notice that **T** is a vector that will be applied on each point, to traslate them to a new location. For instance, look at the picture below.

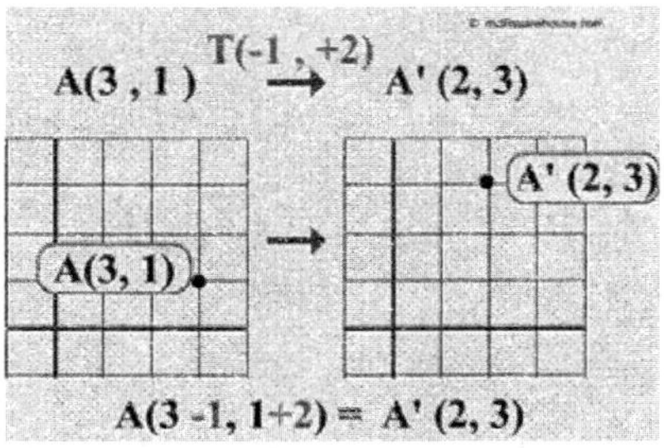

Example: Draw a parallelogram having its vertices A(-5, 2), B(-5, 4), C(-2, 4) and D(-2, 2). After, draw its image under the traslation T(6, -4).

<u>Solution:</u>

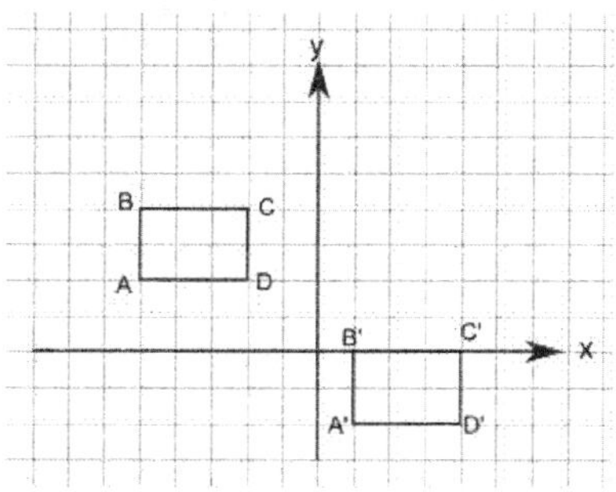

Exercises

16. Traslate the triangle having its vertices A(3,1), B(4,-2) and C(8,-1) by vector **T**(1,4). Do it graphically and analytically and check you obtain the same results.

17. The L-shape A′B′C′D′F′ is the image of the L-shape ABCDEF after which

traslation?

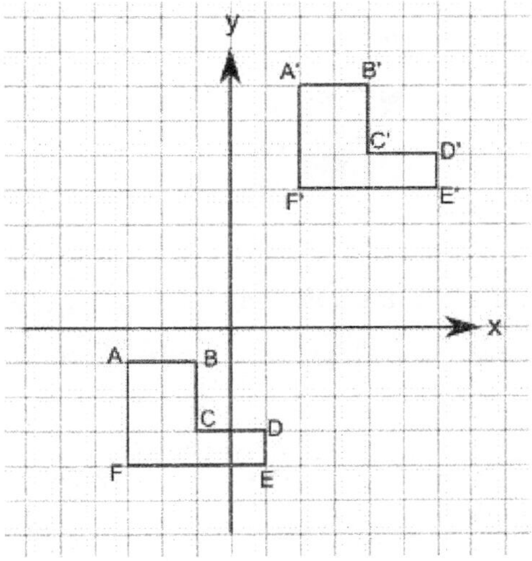

18..In a traslation with vector **T** = (2, 3), we know image of point C under this transformation is C'(-3, 6). Find out the coordinates of point C.

19. In a traslation with vector **T**, point A(1, 0) is transformed into A'(3, 4). Determine coordinates of **T**.

20. Draw the triangle with vertices A(0, 0), B(1, 1) and C(1,0). A traslation **T** = (2, 3) is applied on it. What are the coordinates of transformed triangle?

21. Check that application of two successive traslations with vectors **u** and **v** is another traslation with vector **u** + **v** (this is also named *product* of traslations). Apply to triangle with vertices A(0, 0), B(1, -1) and C(2,4) the traslation of vectors $\mathbf{T}_1$ (3, 2) and $\mathbf{T}_2$ (4, -2).

4. Rotations

Rotation is a very common transformation. Look at the following figures.

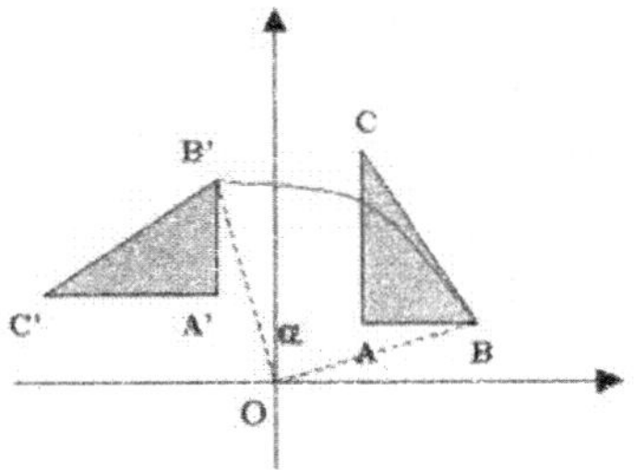

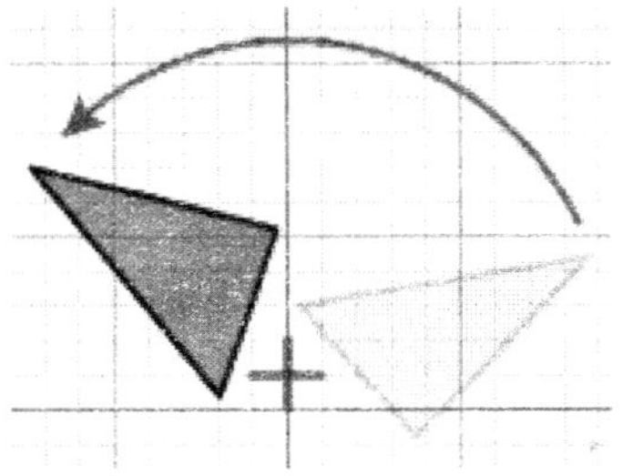

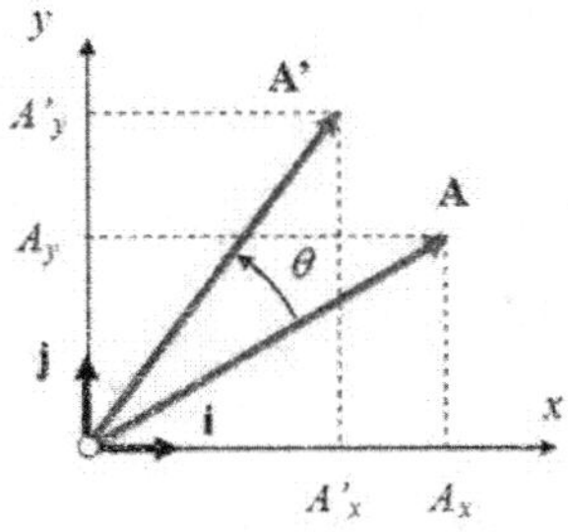

Notation for ***rotation*** is usually ***R(center , degrees)***, where:

- ***Center*** is the 'center of rotation.' This is the point about which you are performing your mathematical rotation.
- ***Degrees*** stands for how many degrees you should rotate. A positive number usually by convention means counter clockwise.

Most common rotations are those ones about the origin. For example R(origin, 90°), a rotation by 90° about the origin can be seen in the picture below in which A is rotated to its image A'. The general rule for a rotation by 90° about the origin is (A,B) → (-B, A). As an exercise build by yourself general rules for rotations by 180°, 270° and 360° about the origin.

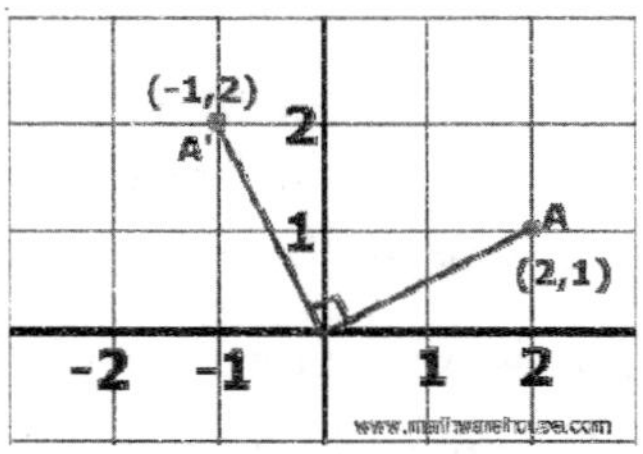

Example: Rectangle A′B′C′D′ is the image of rectangle ABCD after which rotation?

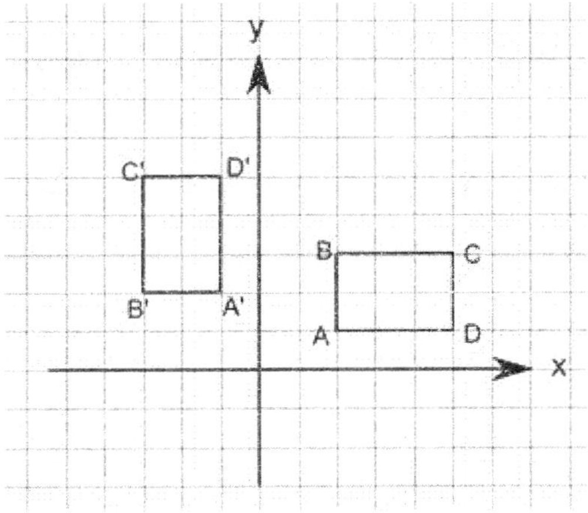

Solution:

Imagine you cut first quadrant and rotate it 90°. You are going to obtain, exactly, image at the second quadrant.

So, rotation has been R(origin, 90°).

Exercises

22. This 'L'-shape (Fig. 1) is rotated about the origin (0,0) by 90° anticlockwise. Draw the new figure

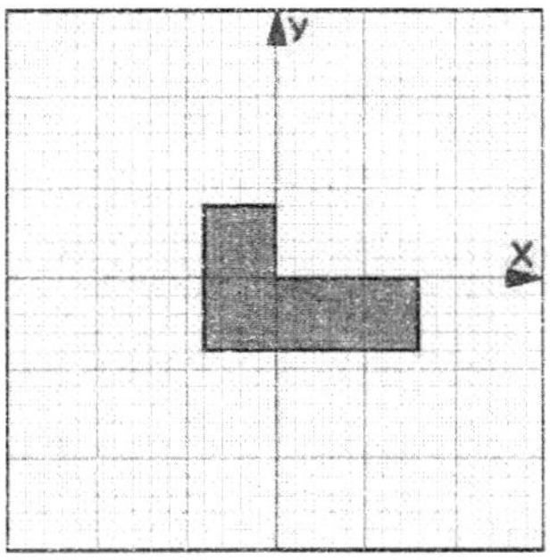

Figure 1

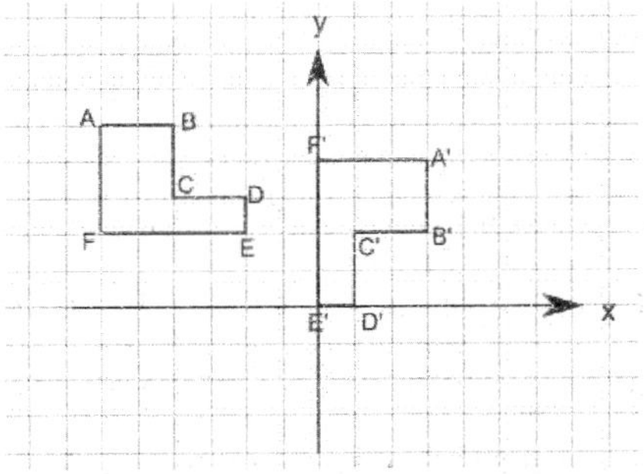

Figure 2

23. The L-shape A′B′C′D′E′F′ (Fig. 2) is the image of the L-shape ABCDEF after which rotation?

24. Draw the triangle with vertices A(-5,0), B(0,5) and C(4,3). Apply rotation **R**(0, 90°).

25. Draw a regular hexagon with center in origin. Apply rotation **R**(0, 60°). What does it happen? And with rotation **R**(0, 120°)?

5. Reflections

Reflection of a point about an axis, named *symmetry axis* is the transformation that relates each point A with another one, A', being A and A', symmetric points about the symmetry axis.

Examples of the most common types of reflections in math are:

Reflection in the x-axis: It can be seen in the picture below in which A is reflected to its image A'. The general rule for a reflection in the x-axis: $(A,B) \to (A, -B)$

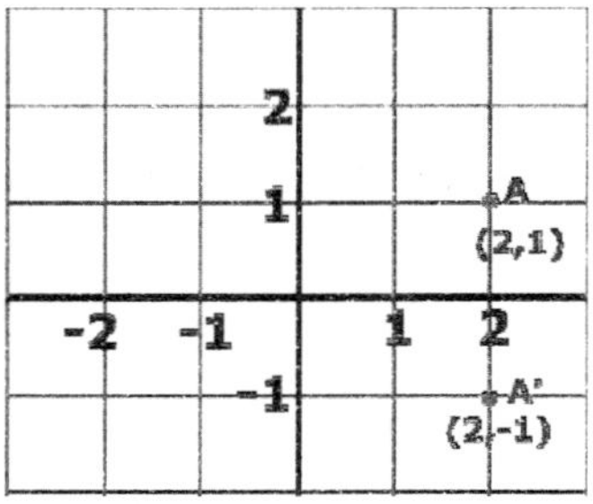

Reflection in the y-axis: It can be seen in the picture below in which A is reflected to its image A'. The general rule for a reflection in the y-axis: $(A,B) \to (-A, B)$

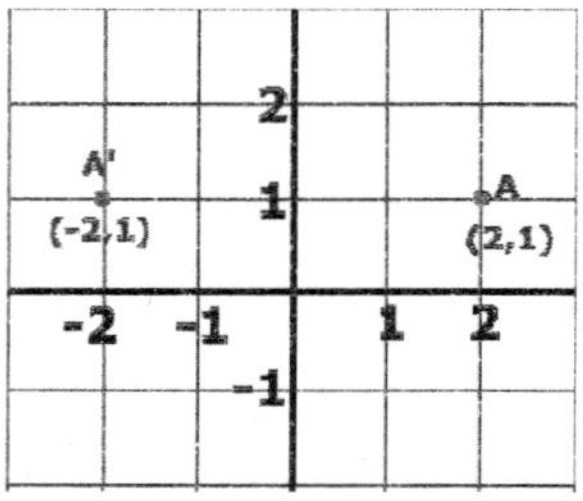

Reflection in the straight line $y = x$: It can be seen in the picture below in which A is reflected to its image A'. The general rule for a reflection in the straight line $y = x$: $(A, B) \to (B, A)$

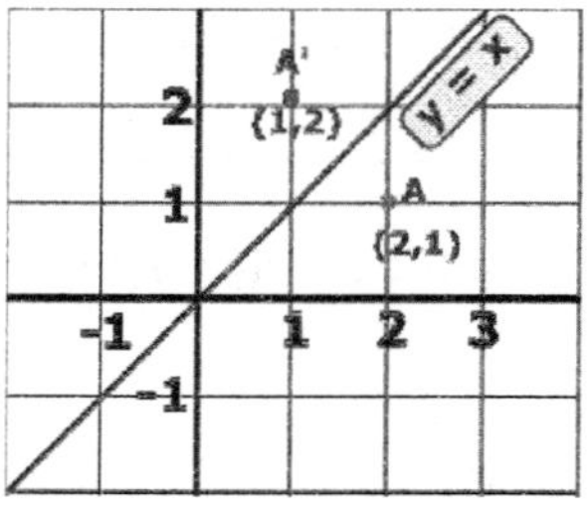

Exercises

26. Rectangle A′B′C′D′ (Fig.1) is the image of rectangle ABCD after a reflection in which straight line?

27. The L-shape A′B′C′D′F′ (Fig. 2) is the image of the L-shape ABCDEF after a reflection in which line?

28. The rectangle in Fig. 3 is reflected in the x-axis and then reflected again in the line y = -3. Draw the final image.

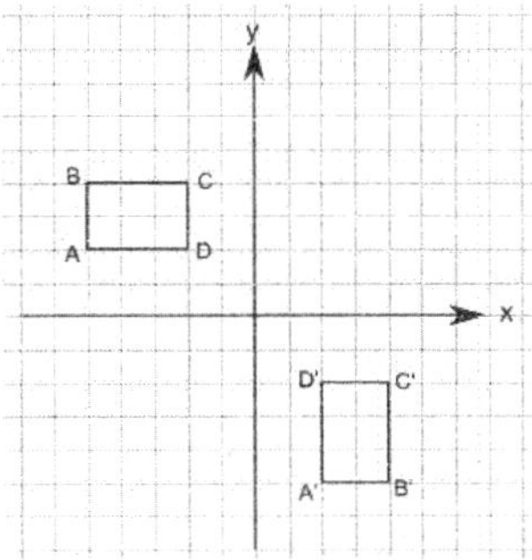

Fig. 1

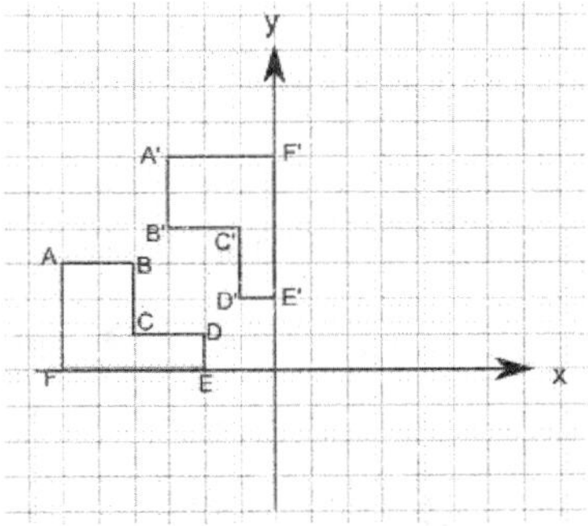

Fig. 2

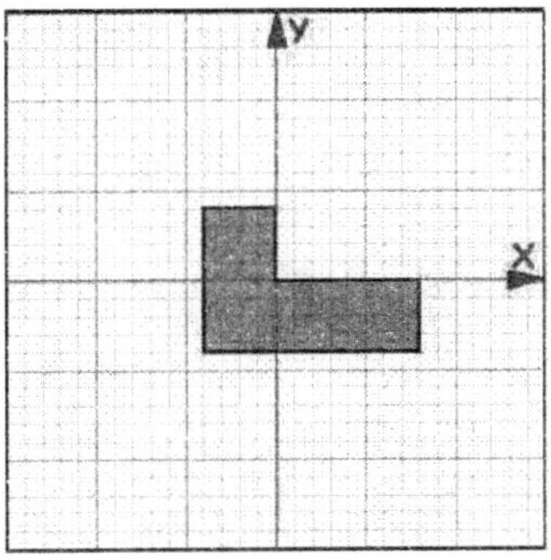

Fig. 3

29. Given the triangle with vertices A(0, 0), B(1, -1) and C(2,4), find out the symmetric triangles about X-axis and Y-axis. Do it analytically. You can draw them at the end, to check your answer.

Review exercises

Vectors

1. Given points A(-3,0), B(3,7), C(5,3), D(-1,-4) and E(-5,4), find out the coordinates of vector.

$\vec{AB}$ $\vec{BC}$ $\vec{AD}$ $\vec{DE}$ $\vec{CD}$ $\vec{BA}$.

2. If point C has coordinates (3,1) and vector $\vec{CD}$ is (-1,4), where is point D located?

3. Subtract vectors $\vec{u} = (2,1)$ *and* $\vec{v} = (-3,4)$.

4. Given vectors $\vec{u} = (-3,1)$ *and* $\vec{v} = (2,4)$ calculate **u+v** and **u-v.**

5. Given vectors $\vec{u} = (1,-3), \vec{v} = (4,2)$ *and* $\vec{w} = (-3,0)$, calculate the coordinates of vectors:

$\vec{u} + \vec{v}$ $\quad$ $\vec{u} + \vec{w}$ $\quad$ $\vec{v} - \vec{w}$ $\quad$ $\vec{u} - \vec{w}$

6. Given vectors $\vec{u} = (-1,3); \vec{v} = (-2,4)$ *and* $\vec{w} = (0,2)$, calculate the following linear combinations:

$a)\ \vec{u} + 2\vec{v}$ $\quad$ $b)\ 2\vec{u} - \vec{v}$ $\quad$ $c)\ \vec{v} + 3\vec{w}$ $\quad$ $d)\ \vec{u} - 2\vec{v} + 3\vec{w}$

7. Given points A(-3,1) and B(2,6), calculate the coordinates of the midpoint of the segment AB.

8. Midpoint of segment AB is M(3,-1). Find out coordinates of point B, knowing A = (2,-4).

Traslations

9. Consider vector **T**(3, 5), having origin in (0, 0) and head in (3, 5).

a) Traslate points A(0, – 4), B(–3, –5), C(0, 0) and D(5, –1) by this vector.

b) Check that points M(1, 3), N(7, –1) and X(4, 1) are on a straight line. Traslate them by vector **T** and check that transformed points are also on a straight line.

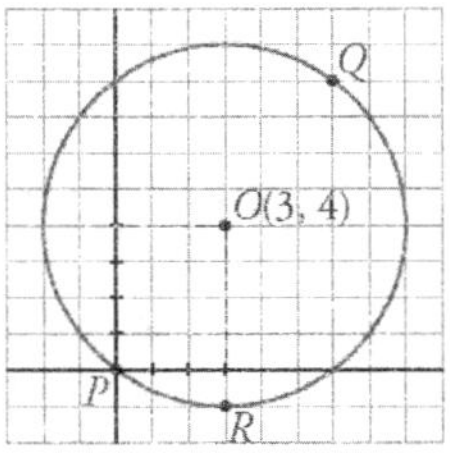

10. Draw axes X and Y in your notebook. Draw the circumference with center C(3,4) and radius 5.

a) Check that circumference passes by *P*(0, 0), *Q*(6, 8) and *R*(3, –1).

b) Traslate points *O*, *P*, *Q* and *R by translation T with vector T*(6, –2).

c) Check that circumference whose center is C' = T(C) and radius 5, passes by *P'*, *Q'* and *R'*.

Rotations

11. Following figures have a rotation center. Indicate the rotation angle in each case.

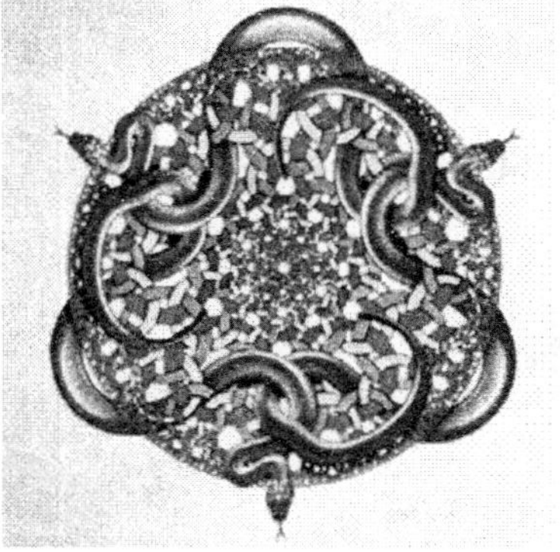

12. In your notebook, draw coordinated axes. Consider the rotation with center 0(0, 0) and angle a = 90°.

a) Transform points $A(-5, 0)$, $B(0, 5)$, $C(4, 3)$ and indicate the triangle $A'B'C'$, transformed of triangle ABC.

b) What is the transformed of straight line r, passing by A and B?

c) What is the transformed of circumference with center O and radius 7?

13. Consider the following mosaic.

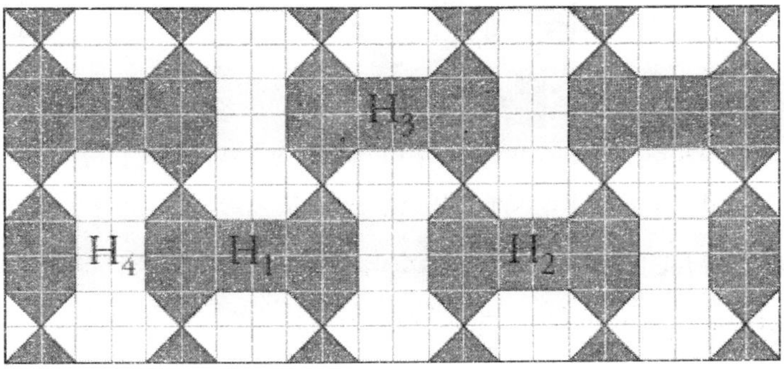

a) Describe rotation that transforms H_1 into H_4.

b) Describe rotation that transforms H_1 into H_3.

14. Find the symmetry axes of the following figure:

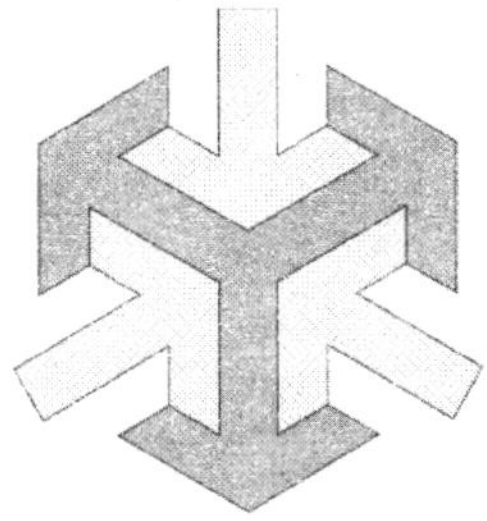

15. We are going to consider the symmetry **S** of the straight line $y = x$. Draw the transformed under **S** of:

a) Points $A(3, 1)$, $B(4, 0)$, $C(0, 4)$, $D(5, 5)$.

b) *X-axis.*

c) *Y-axis.*

d) Circumference with center (1, 4) and radius 2.

e) Circumference with center (3, 3) and radius 5.

16. In your notebook, draw triangle Δ with vertex $A(-5, 3)$, $B(-2, 2)$, $C(0, 5)$. Consider the traslation

T(5, −1) and the reflection about the X-axis.

a) Transform Δ by **T**x**S**. b) Transform Δ by **S**x**T**.

17. In your notebook, complete the following mosaics:

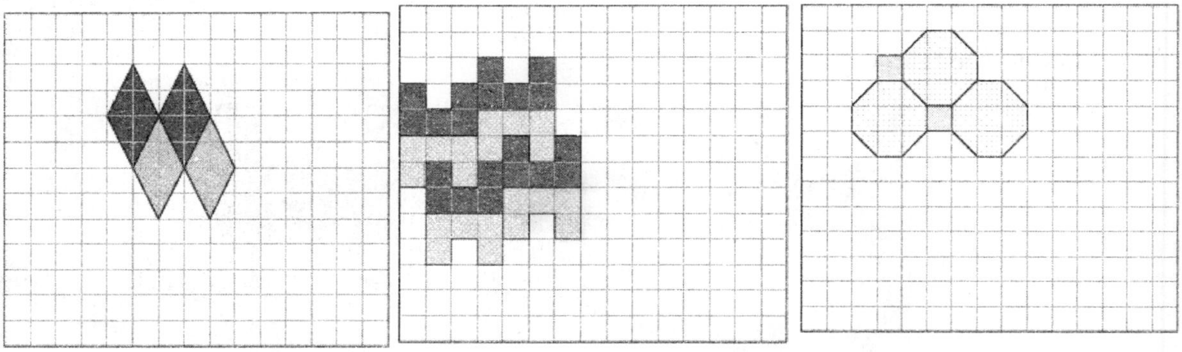

Unit 11.- Solids

1. Polyhedrons

In geometry, the word ***solid*** refers to the shell of a three-dimensional figure, not including its interior. Solids can be classified in many ways, based on their shape.

- **Polyhedron:** It is the solid delimited by polygons.

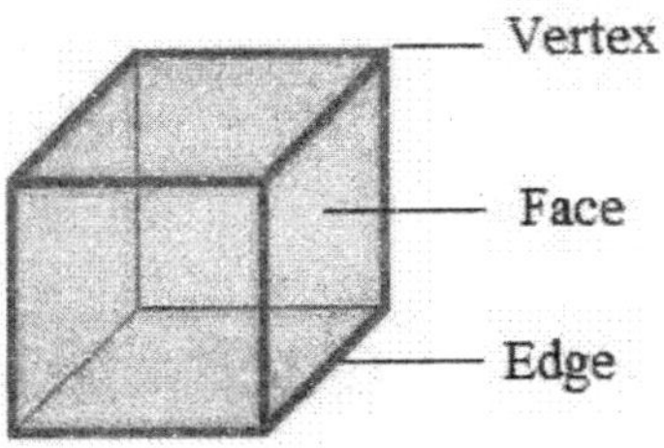

The parts of a polyhedron are:

- **Face:** Each polygon that forms the polyhedron.
- **Edge:** Intersection of two faces.
- **Vertex:** Intersection of two or more edges.
- **Vertex order:** Number of faces intersecting in a vertex.

Polyhedrons can be classified under different criteria:

a) Faces being identical regular polygons or not.

b) Dihedral angles.

c) Number and shapes of their faces.

a) Regular and irregular polyhedrons: *Regular* polyhedron is that one in which all its faces are identical regular polygons and all its vertices have the same order. The rest of the polyhedrons are named *irregular*.

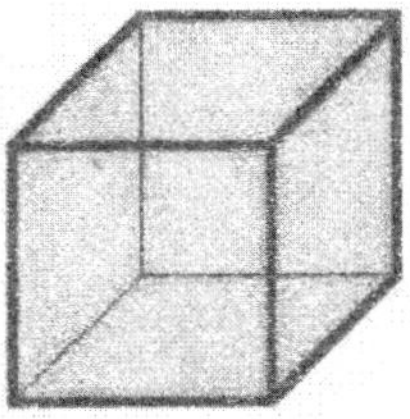

Regular

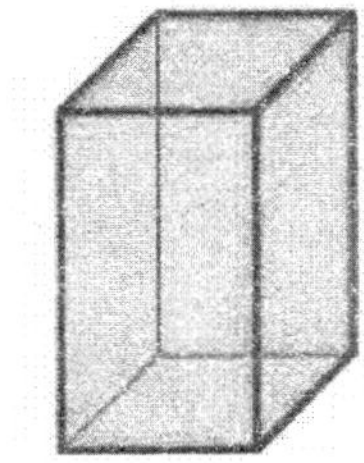

Irregular

There are only five convex *regular polyhedrons*, and they are known collectively as the ***Platonic solids***, shown below.

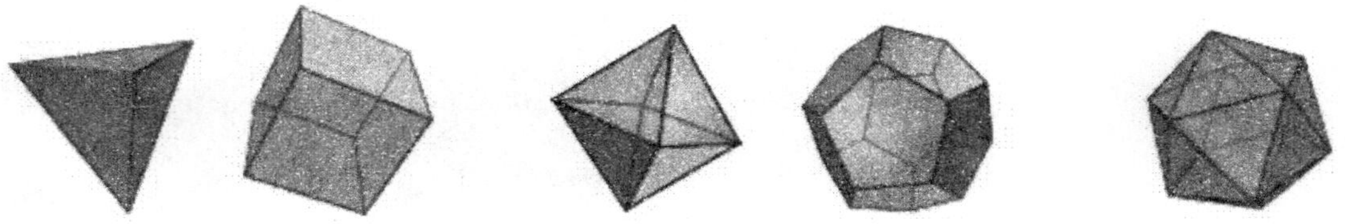

TETRAHEDRON (4 faces) HEXAHEDRON (Cube) (6 faces) OCTAHEDRON (8 faces) DODECAHEDRON (12 faces) ICOSAHEDRON (20 faces)

b) Convex and concave polyhedrons: A *convex* polyhedron is that in which all the dihedral angles are less than 180°. On the other hand, a *concave* polyhedron is that one that has, at least, one dihedral angle bigger than 180°.

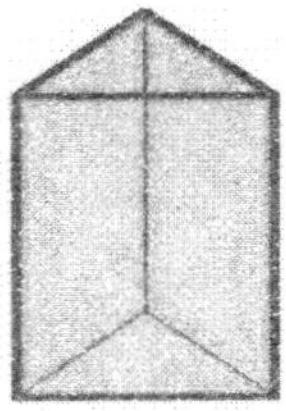

Convex

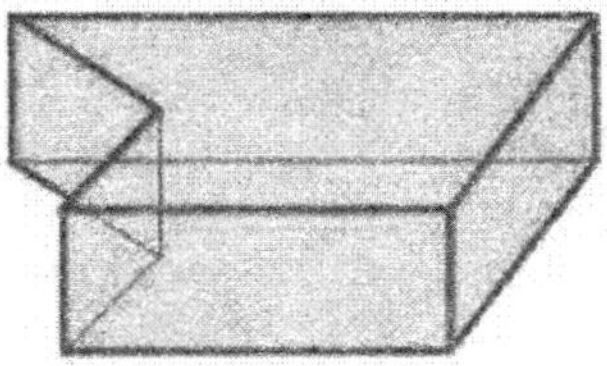

Concave

Exercises

1. Develop a 2 cm side octahedron.

2. Develop a 4 cm side cube.

3. Draw an oblique prism with a triangular base.

1.1. Prisms

A prism is a polyhedral with two identical parallel bases, whose lateral faces are parallelograms.

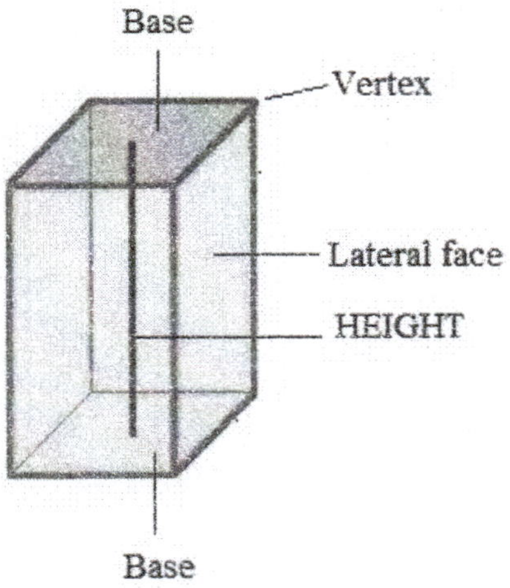

- The **height** is the perpendicular distance between the bases.
- Prisms are classified depending of the polygon in their bases, into *triangular*, *quadrangular*, etc.
- **Right**: All its lateral faces are rectangles (they are also perpendicular to the bases).
- **Oblique:** Not all its faces are rectangles (they are not perpendicular to the bases).

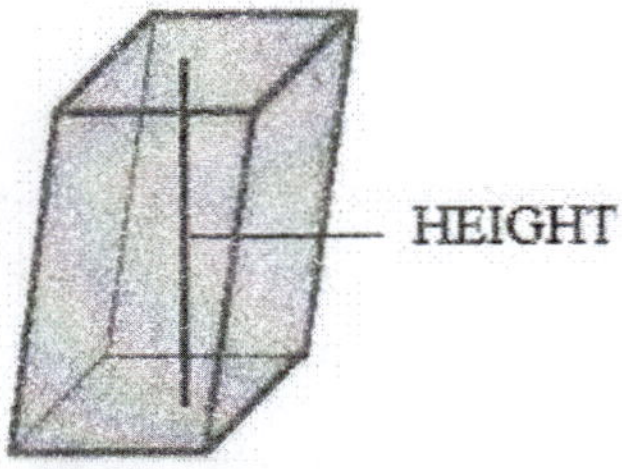

Oblique prism

$A_{prism} = 2 \cdot A_{Base} + A_{lat}$

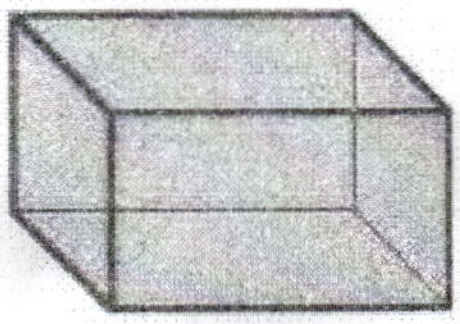

Parallelepiped

Orthohedron

- **Parallelepiped**: Prism whose bases are parallelograms.

- **Orthohedron**: Parallelepiped in which 6 faces re rectangles.

- **Regular Prisms**: Those that are right and their bases are regular polygons.

1.2. Pyramids

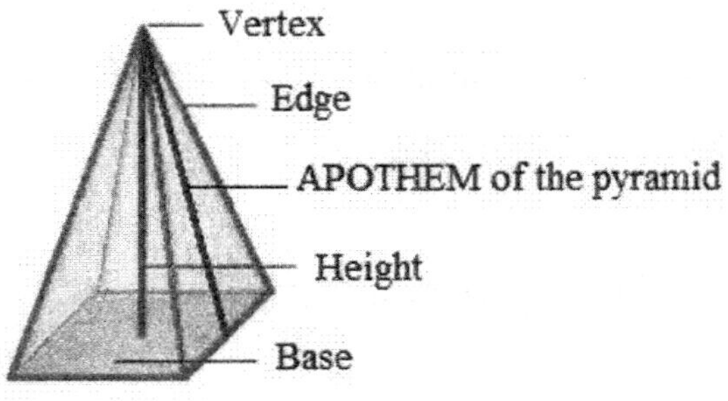

A ***pyramid*** is the polyhedron whose base is a polygon and its lateral faces are triangles converging in the vertex of the pyramid.

- **Height:** distance from the vertex to the base.

- **Right pyramid:** A pyramid whose lateral faces are isosceles triangles (Also: the height begins in the center off the base).

- **Oblique pyramid:** It is nor a right one.

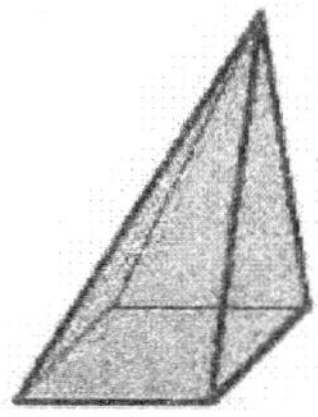

Oblique pyramid

- **Regular pyramid:** a right pyramid having in the base a regular polygon.

- **Apothem of a regular pyramid:** height one of its faces.

Pyramids can be classified depending on the type of polygon having in their base, as *triangular*, *quadrangular*, etc.

Exercises

4. What is the polygon in the base of a pyramid if it has 5 lateral faces?

5. What is the polygon in the base of a pyramid that has 8 faces?

6. Draw the development of a pyramid having all its face identical.

7. Which of the following figures is the development of a pyramid?

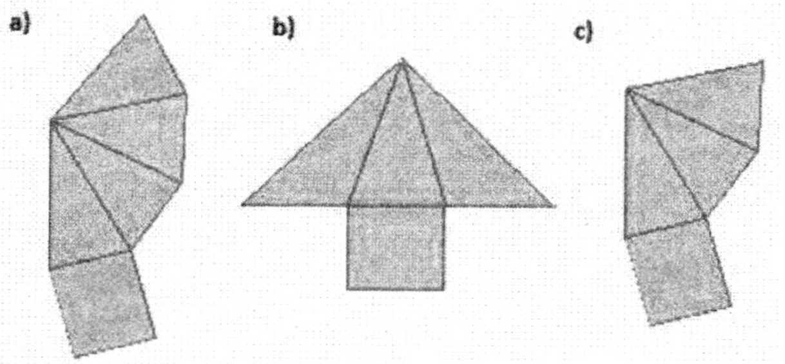

2. Solids of revolution

Round solids or ***solids of revolution*** are solids that are generated by rotating a plane shape about one of its sides. So, they have only **one** lateral face, which is curved.

2.1. Right Cylinder

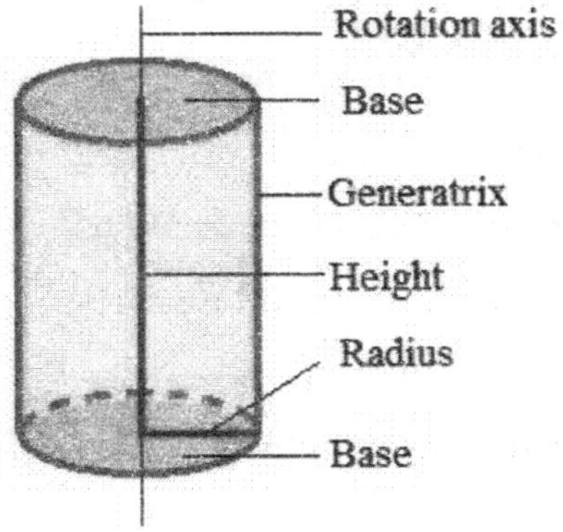

A ***right cylinder*** is the solid generated by rotating a rectangle about one of its sides.

- **Rotation axis:** side about which rectangle has rotated.

- **Generatrix:** side of the rectangle that is opposite the rotation axis.

- **Height:** (perpendicular) distance between the bases.

- Its bases are circles.

- **Radius** of the base is one of the perpendicular sides to the rotation axis.

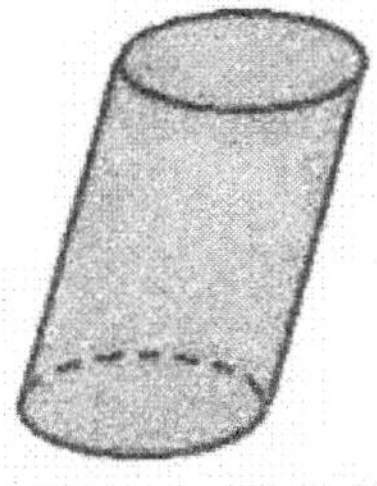

Oblique cylinder

- **Oblique cylinder:** it is the cylinder resulting of cutting a right cylinder by two parallel planes that are not perpendicular to the rotation axis.

2.2. Right cone

A ***right cone*** is the solid that is generated by rotating a right triangle about one of its legs.

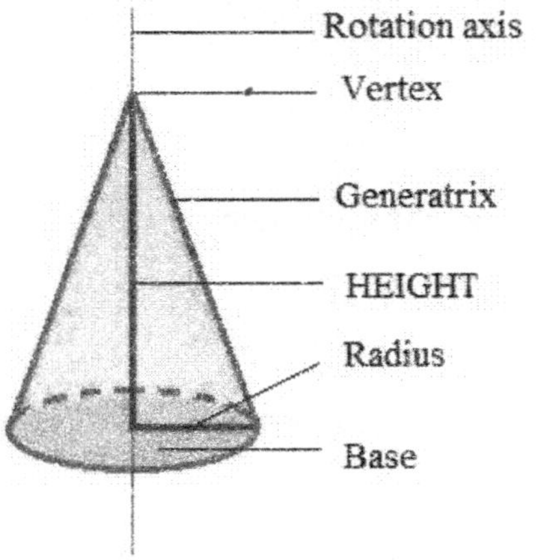

- It has one **base** and one **vertex**.

- **Rotation axis:** leg about which right triangle rotates.

- **Generatrix:** hypotenuse of the right triangle.

- **Height:** Segment from the vertex to the base. It is the rotation axis.

- **Radius** of the base is the leg of the right triangle that is¡s not the rotation axis.

2.3. Sphere

A ***sphere*** is the solid that is generated by rotating a semicircle about its diameter.

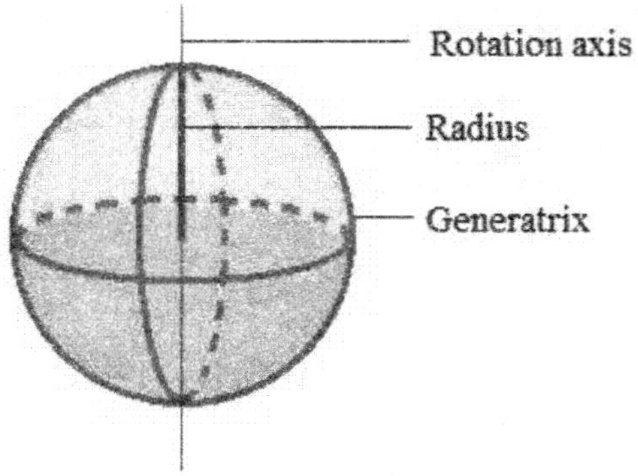

- **Rotation axis:** it is the semicircle diameter.

- **Generatrix:** it is the semi-circumference of the semicircle.

- The **radius** of the sphere is the semicircle radius.

Exercises

8. Classify the following solids into polyhedrons or solids of revolution. For solid of revolution, indicate the plane shaped figure that generates them:

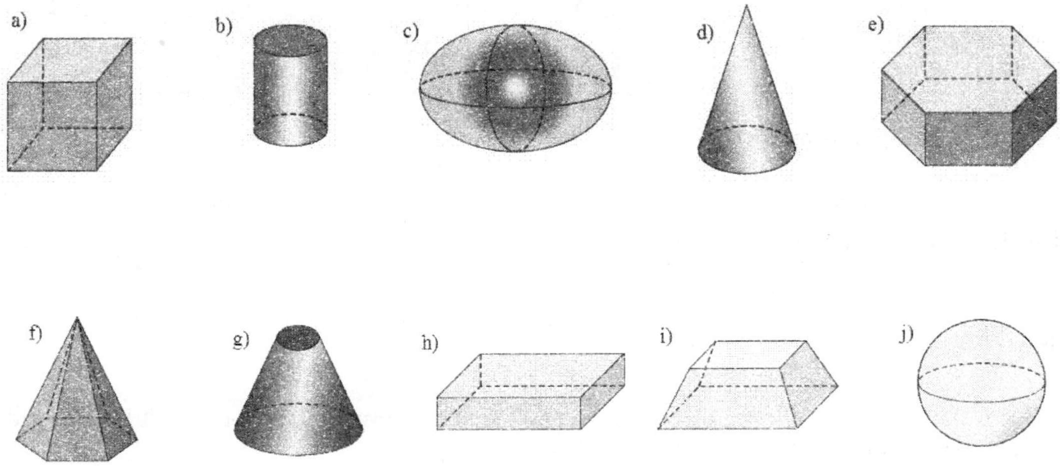

9. Draw the solid of revolution generated when the figure given below (Figure 1) rotates about its left lateral segment.

10. What solid is generated when rotating the given trapezium (Figure 2) about its right side?

11. What solid is generated when rotating the given trapezium (Figure 3) about its left side?

12. What solid is generated when rotating the given trapezium (Figure 3) about its right side?

13. What solid is generated when rotating the given trapezium (Figure 4) about its right side?

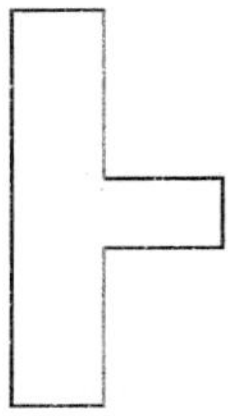

Figure 1

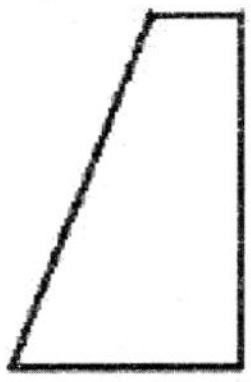

Figure 2

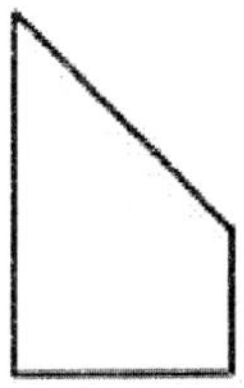

Figure 3

Figure 4

3. Symmetry planes of a solid

A ***symmetry plane*** of a solid is a plane that divides it into two halves that are symmetric about that plane.

Exercises

14. a) What conditions must a plane verify to be a symmetry plane of a tetrahedron?

b) How many symmetry planes does a tetrahedron have?

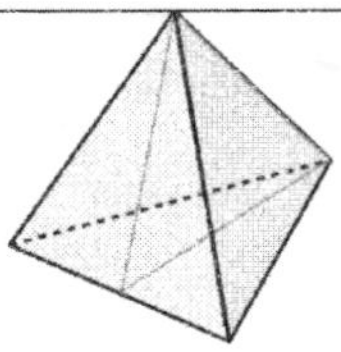

15. Draw a regular hexagonal prism. How many symmetry planes does it have? And a regular hexagonal pyramid?

16. How many symmetry planes do a cone and a sphere have?

4. Rotation axis of a solid

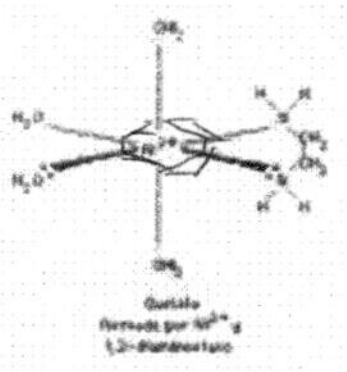

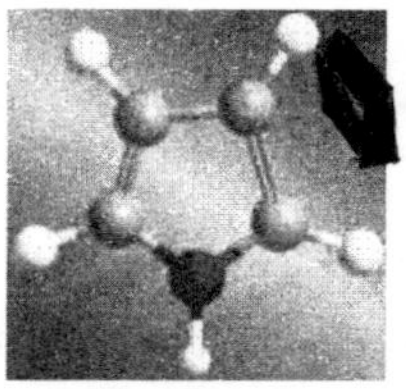

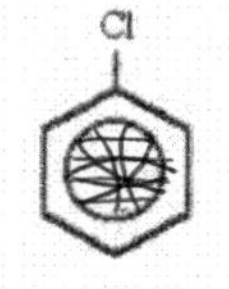

A straight line, *r*, is said to be a ***n-order rotation axis*** of a solid if when rotating the figure about *r* for a whole round, we obtain *n* equivalent positions.

Exercises

17. a) What rotation axes does a regular hexagonal pyramid have? What are their order?

b) And a regular hexagonal prism?

5. Areas of solids

Just to remember:

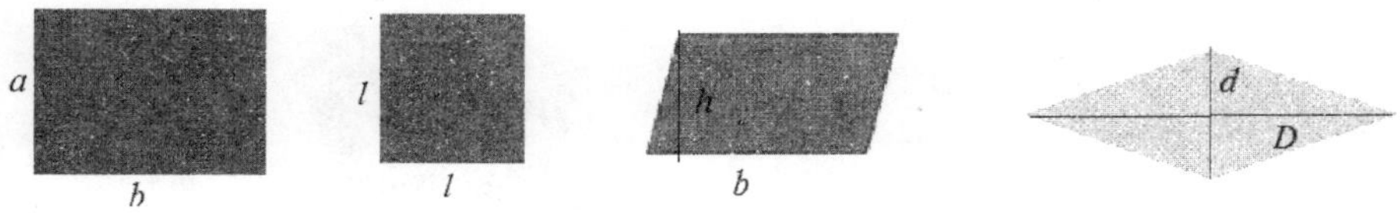

$A = b \cdot a$ $\qquad$ $A = l^2$ $\qquad$ $A = b \cdot a$ $\qquad$ $A = \dfrac{D \cdot d}{2}$

$A = \dfrac{b + b'}{2} \cdot h$ $\qquad$ $A = \dfrac{Per \cdot ap}{2}$ $\qquad$ $A = \dfrac{b \cdot h}{2}$ $\qquad$ $A = \dfrac{b \cdot a}{2}$

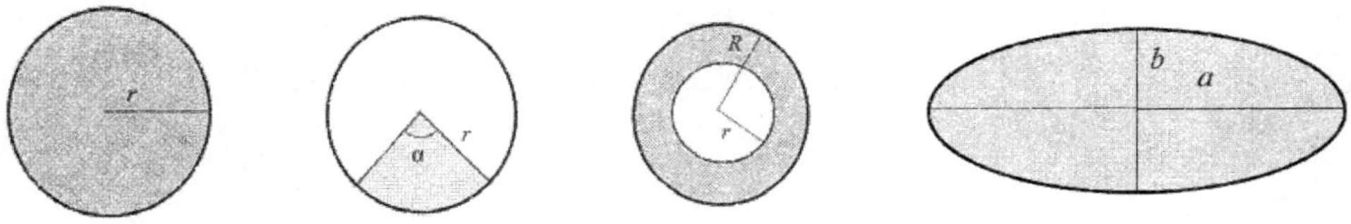

$A = \pi \cdot r^2$ $\qquad$ $A = \pi \cdot r^2 \cdot \dfrac{\alpha}{360}$ $\qquad$ $A = \pi \cdot (R^2 - r^2)$ $\qquad$ $A = \pi \cdot a \cdot b$

And two important solids: Sphere and Cone:

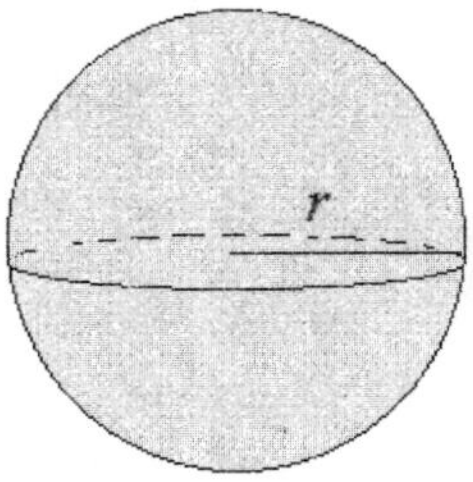

$A = 4 \cdot \pi \cdot r^2$

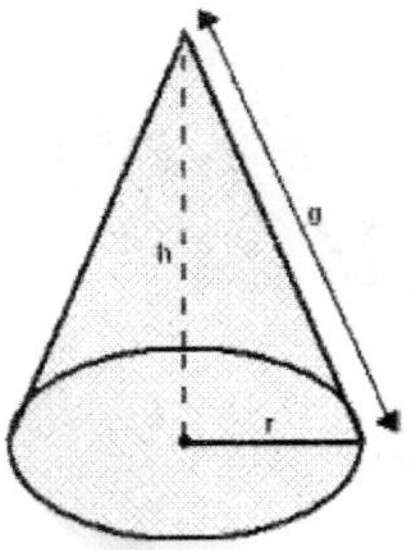

$A = \pi \cdot r \cdot (g + r)$

Exercises

18. Calculate the area of these polyhedrons, obtained from cubes of edge 12 cm.

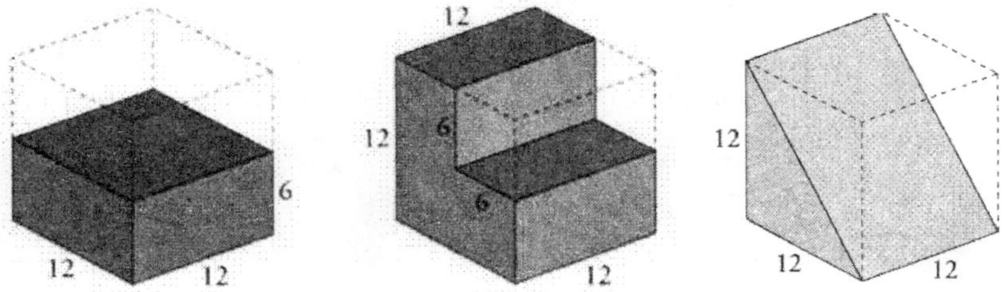

19. Calculate the area of the prism and the pyramid. In both cases, base is a regular hexagon.

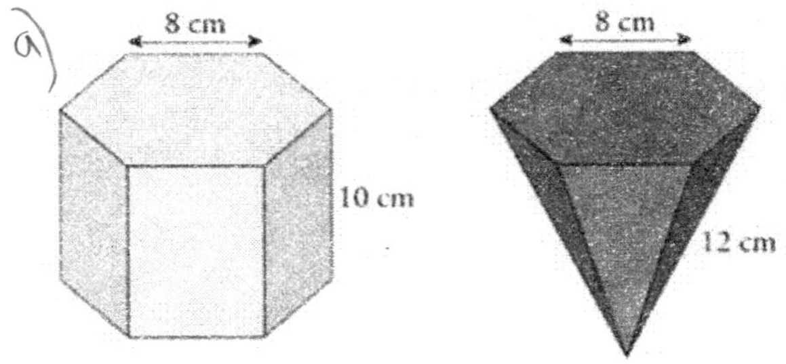

20. Calculate the area of these solids:

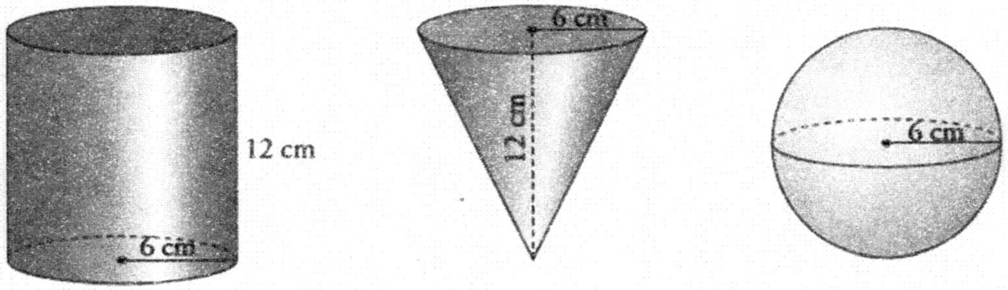

21. Calculate the area of these solids.

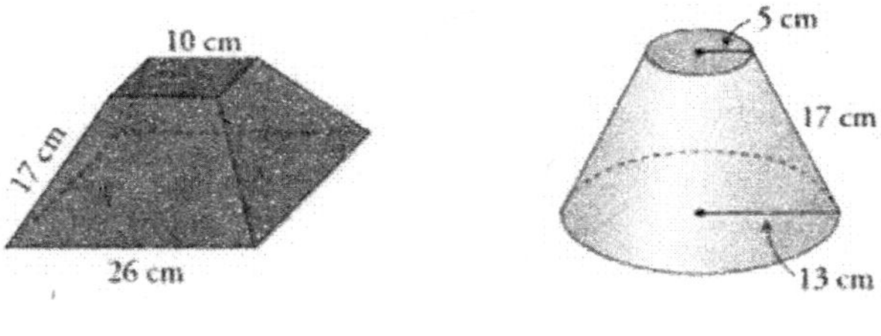

22. Calculate the areas of:

a) A right prism whose base is a rhombus of diagonals 12 cm and 20 cm, knowing that its lateral edge measures 24 cm.

b) A right pyramid with the same base and edge that previous prism.

6. Volume of solids

In order to organize correctly this part of the unit and make it easier to you, we are going to divide solids into three groups: right, appointed and curved solids.

- **Having two bases:** Those having two identical parallel bases. Their representatives are prism and cylinder.

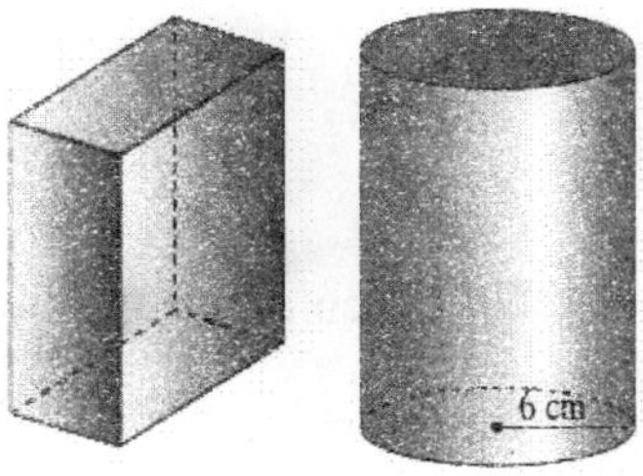

- **Having 1 base and 1 upper vertex:** Those having a base and an appointed head. Their representatives are pyramid and cone.

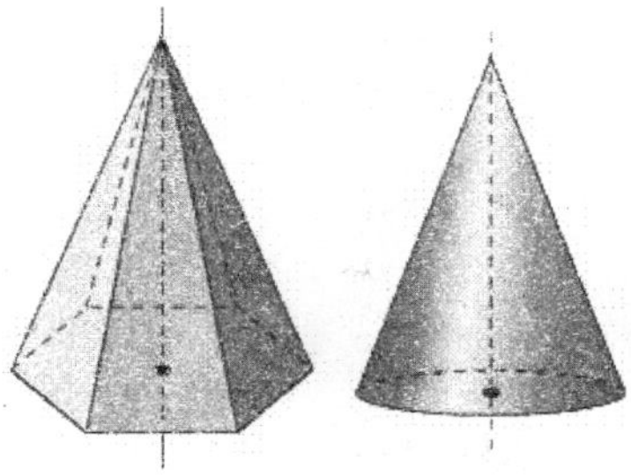

- **Curved solids:** Sphere and those derived from sphere.

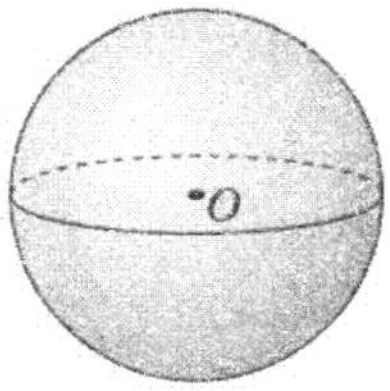

Corresponding solids: As you can see, a pyramid can be imagined into a prism, or a cone into a cylinder. We are going to name these pairs of solids, *corresponding solids*.

Now, study of solids' volume is going to be very systematical:

SOLID	VOLUME
2 bases	$V = A_{base} \cdot h$
1 base, 1 vertex	$V = \frac{V_{CORRESPONDING}}{3}$ or $V = \frac{A_{base} \cdot h}{3}$
Sphere	$V = \frac{4}{3} \cdot \pi \cdot R^3$

Exercises

23. Calculate the volume of these solids, obtained from prisms.

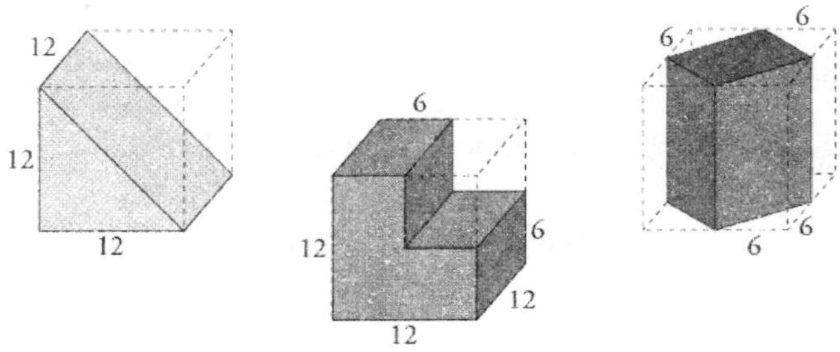

24. Calculate the volume of these solids, whose bases are regular polygons.

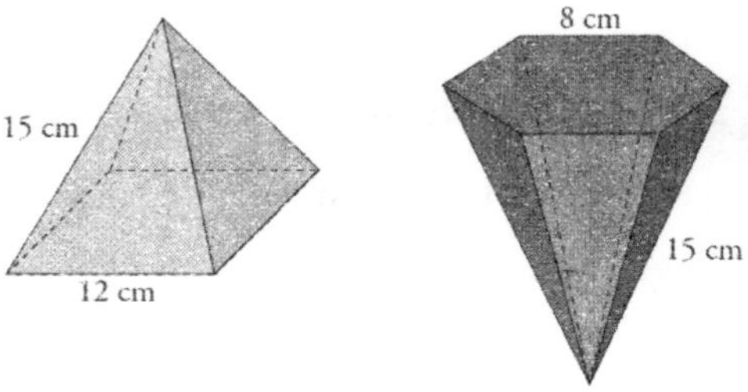

25. Calculate the volume of these solids.

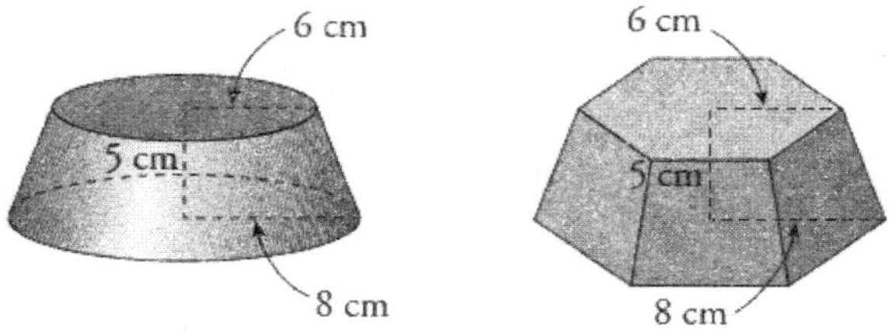

Review exercises

1. Can the figure given below be the development of a cylinder?

2. If we build a cylinder with a rectangular paper sheet, will we obtain the same cylinder by rolling it by its base or by its height?

3. Draw the development of a cone with radius of the base of 5 cm and generatrix of 10 cm.

4. With a 4 cm base and 8 cm height triangle we are generating a cone by rotating it about its height. What is the length of its generatrix?

5. Can the development of a cone be a whole circle?

6. When rotating a quarter of a circle about a radius, what solid will we obtain?

7. When rotating a circle about an exterior axis, what solid are we obtaining?

8. Draw the following solids and calculate their volumes:

a) A cube with edge 9 m. Calculate, also, its area.

b) A regular triangular right prism with basic edge 5 cm and height 16.5 cm. Calculate, also, its area.

c) A regular hexagonal right prism with basic edge 8 cm and height 10 cm. Also, its area.

d) A right cylinder with radius 3 cm and height 10 cm.

e) An oblique circular cylinder with radius 3 mm and height 5 mm.

f) A right cone with height 4 cm and base radius 3 cm.

9. Calculate the volume existing between the cube and the cone:

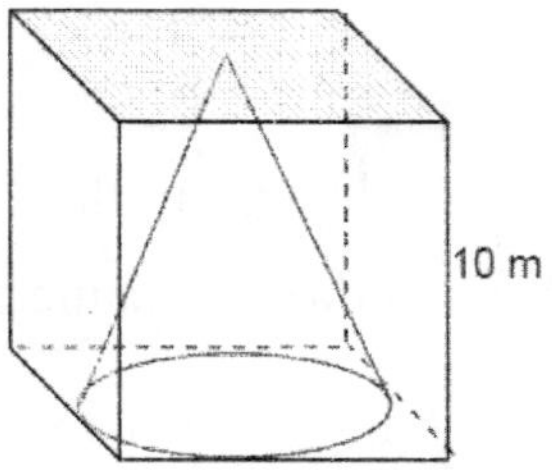

10. Calculate the area of this figure:

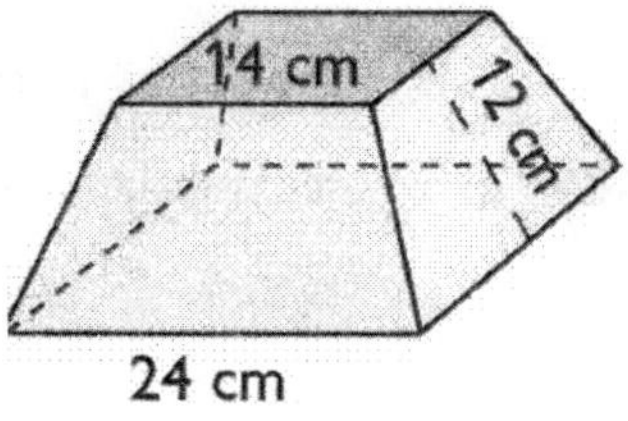

11. Calculate area and volume of first figure, and volume of second one:

a)

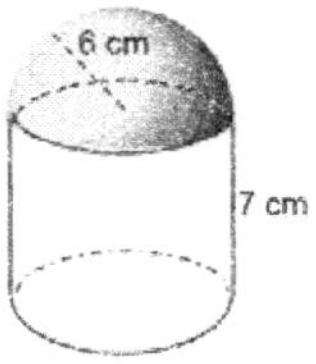

b)

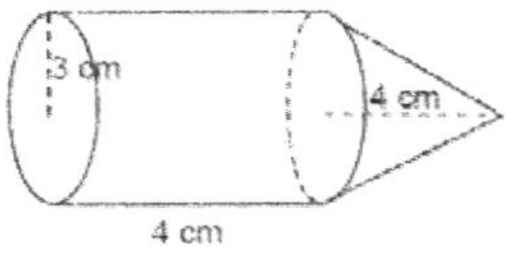

12. a) Estimate the volume and the area of the Earth, knowing its radius is, approximately, 6371 km.

b) Atmosphere has a mean height of 100 km. Estimate its volume and the percent of the total volume that it represents.

13. Calculate volume of Kio Towers, in Madrid, knowing its base is a square of side 35 m and its height is 114 m.

14. We want to paint walls and ceiling of a dining-room whose floor is 12 x 7 m, and whose height is 3.5 m. This room has two doors with dimensions 1 x 2 m, and three windows with dimensions 2 x 2 m.

a) How much area is there to paint? (Draw it).

b) If we have painting cans for 25 m^2, how many cans do we need?

15. Calculate the volume of a Rubik's cube with edge 8 cm.

16. Calculate the volume, in ml, of a can of Coca-cola, whose height is 10.9 cm and whose diameter is 6.2 cm. (1 ml = 1 cm^3)

17. Calculate the volume of Keops' pyramid, knowing its height is 230.35 m and its base is a square of side 136.86 m.

18. A water deposit is a prism whose height is 10 m and whose volume is 4000 m^3. Calculate the side of the base, knowing it is a square.

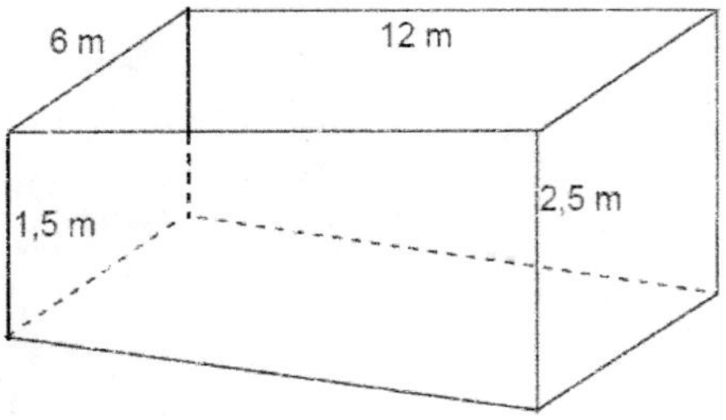

(EXERCISE 19)

19. Calculate the volume of the shown swimming pool. ($1\ m^3 = 1\ 000\ l$)

20. About previous exercise, how long would it take to fill it if water comes at 0.5 l/s?

21. Base diameter of a cylinder is equal to its height. Total area is $169.56\ m^2$. Calculate its dimensions.

22. Calculate the area of the following solids:

a)

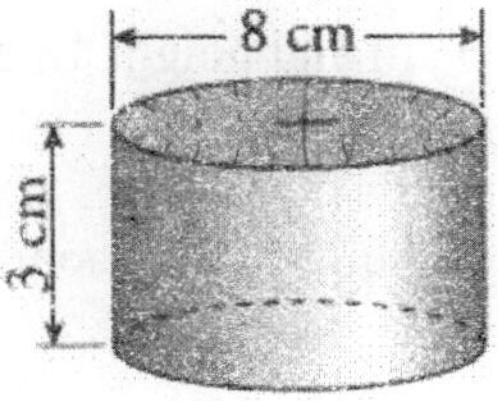

b)

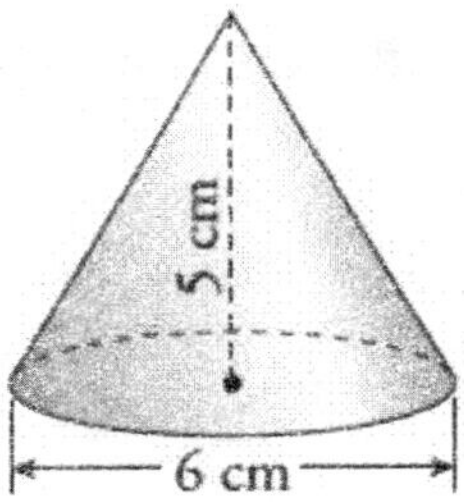

c)

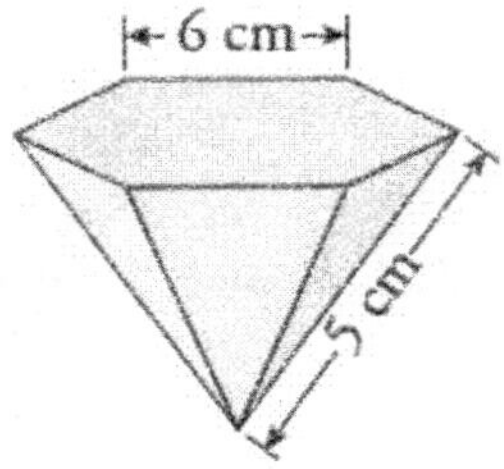

d)

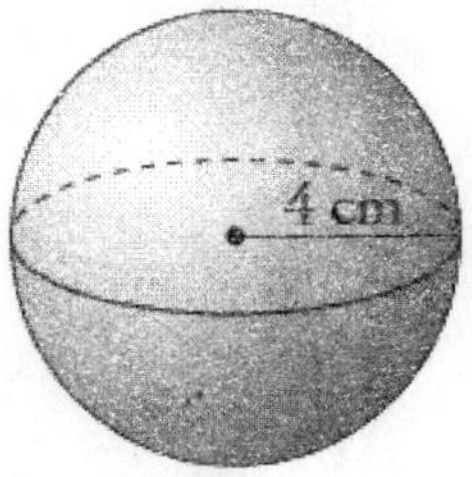

e)

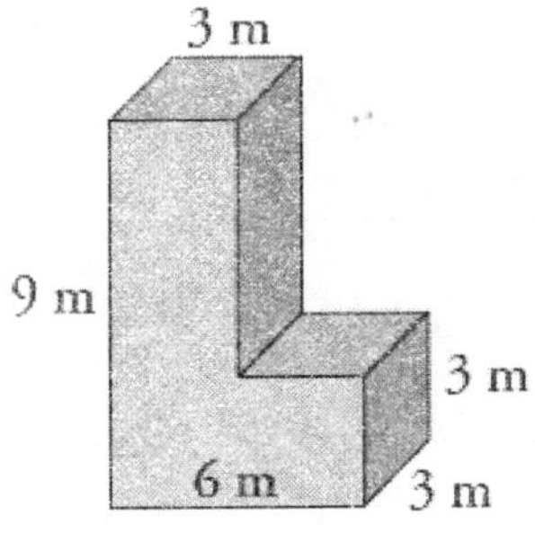

f)

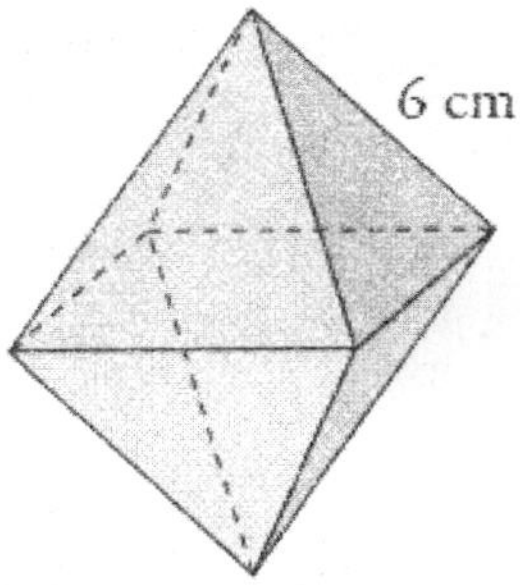

23. Calculate the volume of the following solids:

a)

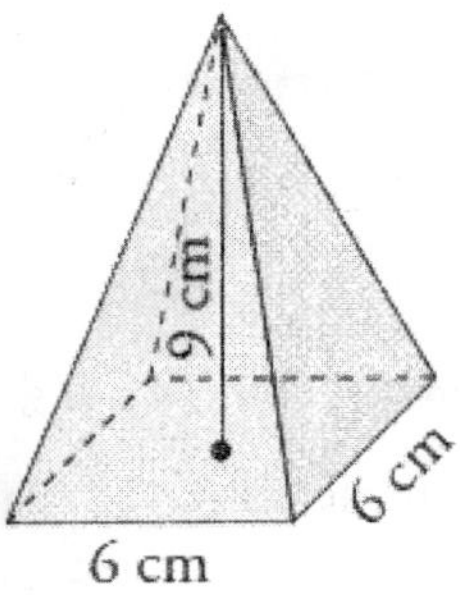

b)

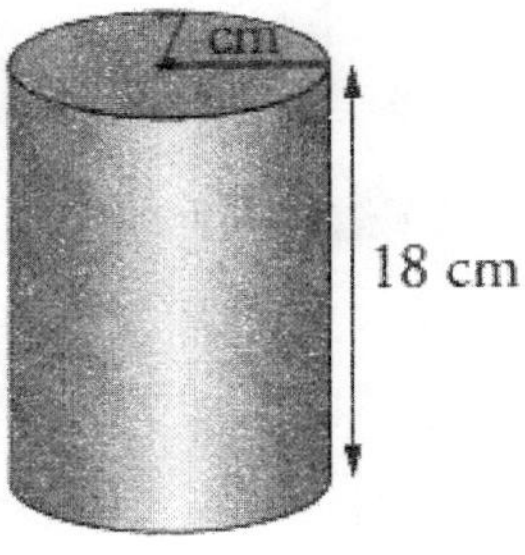

c)

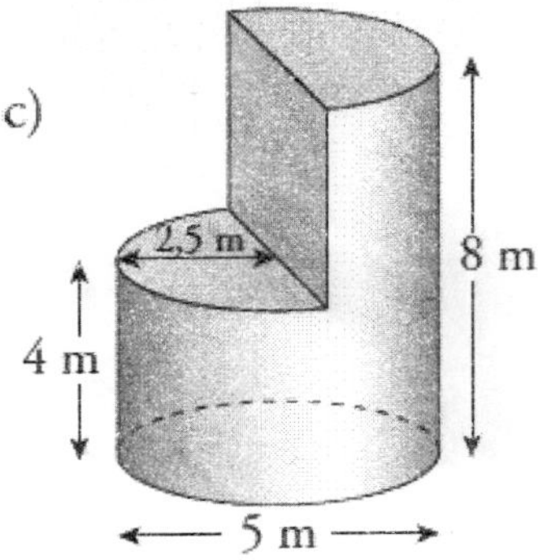

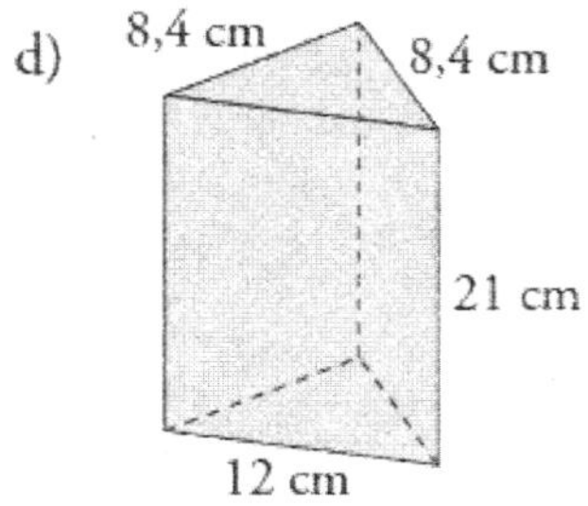

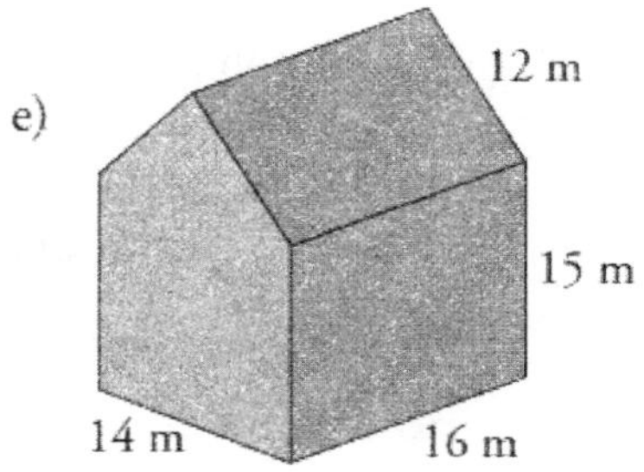

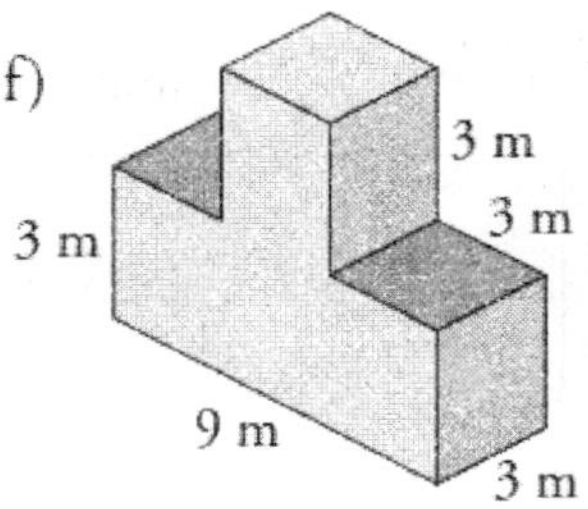

24. Calculate the volume of following solids:

a) Regular octahedron with edge 10 cm.

b) Regular hexagonal pyramid with a lateral edge of 15 cm and basic edge of 8 cm.

c) Semi-sphere with radius 10 cm.

d) Cylinder inscribed in a right prism with a square base with side 6 cm and height 18 cm.

25. Calculate the volume of this regular tetrahedron. Clue: To calculate height H, remember that $\overline{AO} = \frac{2}{3}h$, where h is the height of a face.

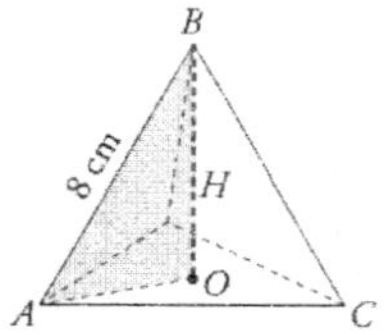

26. Calculate the volume of following solids:

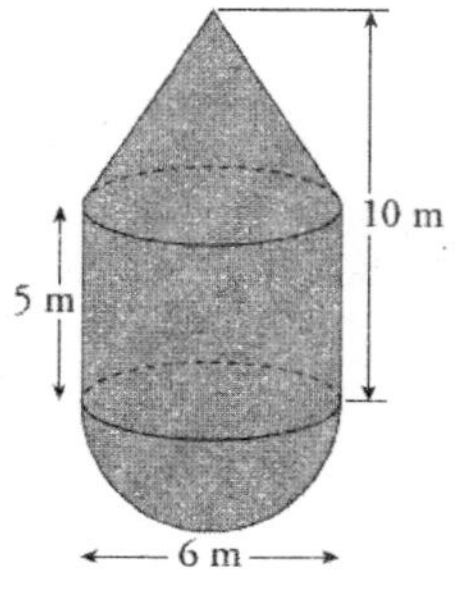

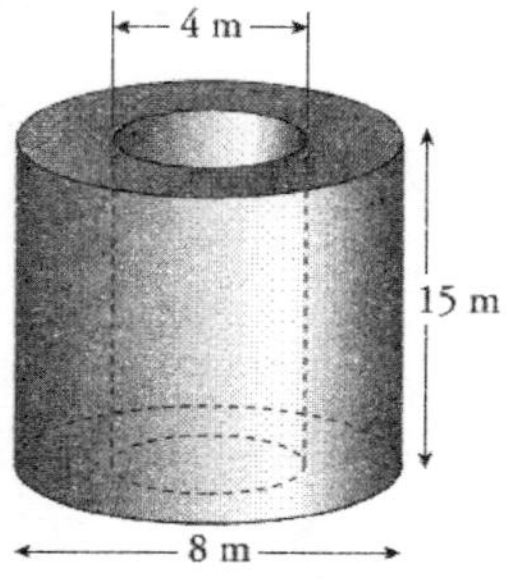

27. Calculate the volume of this solid. Look at the drawing at the right of the solid for help. Units are in meters.

28. Which of these glasses has a higher volume?

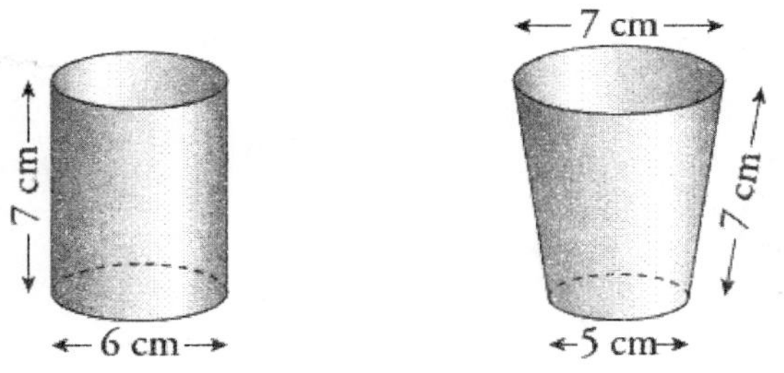

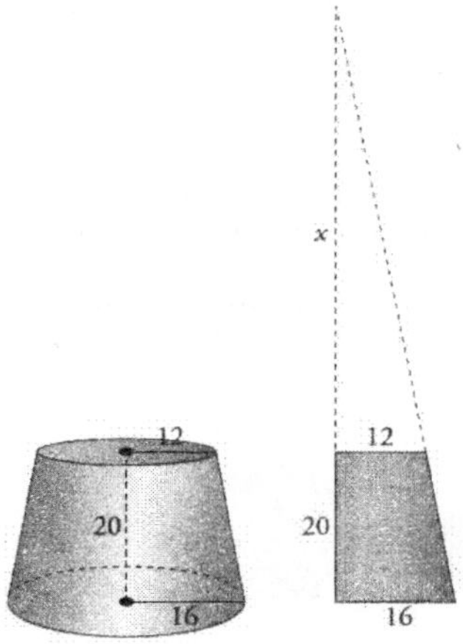

(EXERCISE 27)

29. What is the area of the coloured triangle?

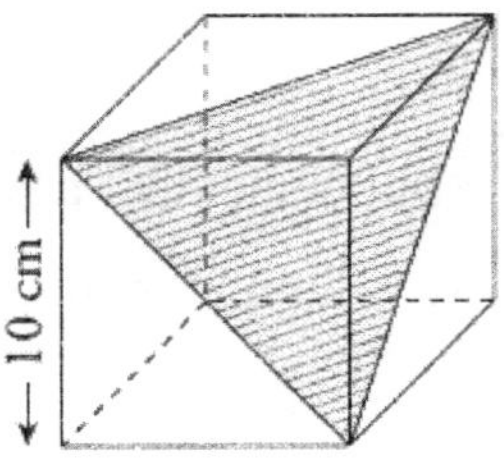

30. What is the volume of the room whose floor is shown?

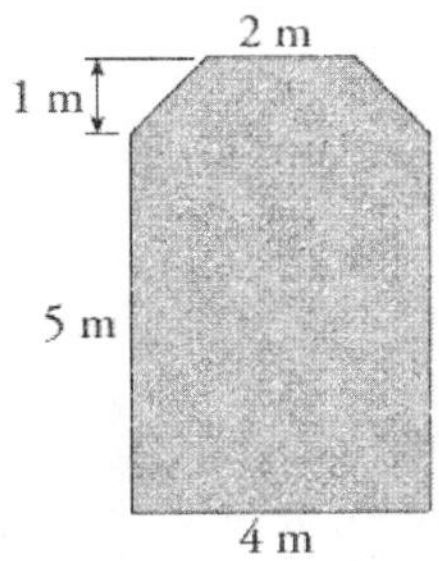

31. Calculate the volume of the solids generated when rotating these plane shapes around the indicated axis.

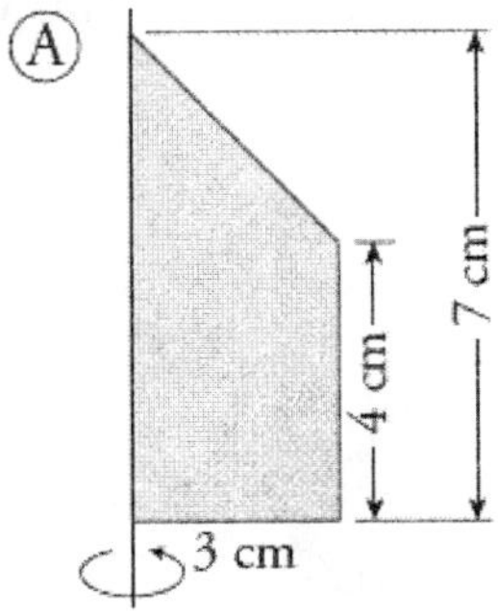

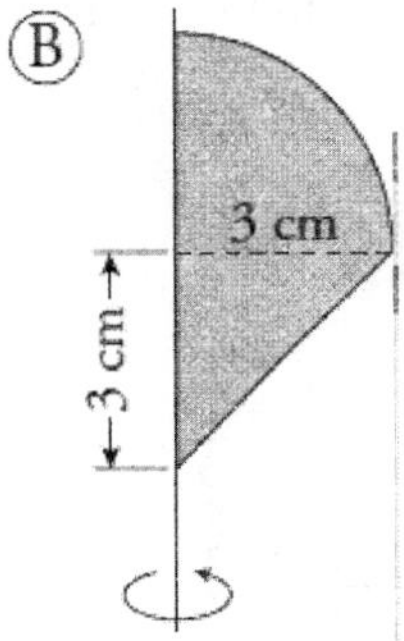

32. Three tennis balls are kept in a cylindrical box with 6.6 cm diameter. Calculate the volume of empty space.

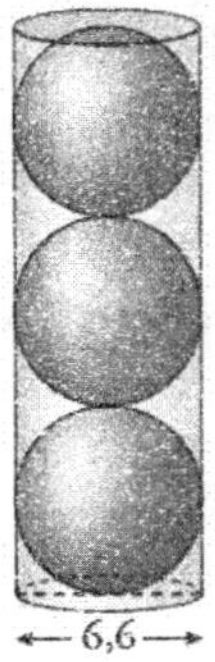

33. We put a 14 cm diameter ball in a 14 cm edge cube, initially full of water. Calculate: a) Volume of dropped water. b) Height of water remaining in the cube.

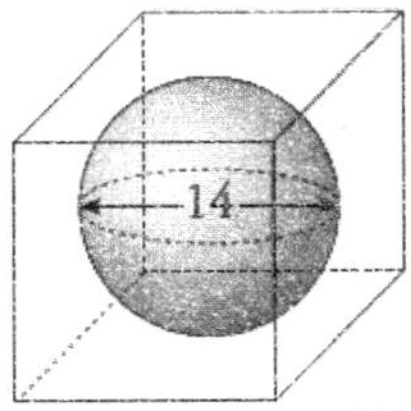

34. A isosceles right triangle has legs of 8 cm. Calcuate the volume of the solid generated when rotating around its hypotenuse.

35. We want to build a cylinder by joining parallel sides of a rectangle whose sides measure 20 and 28 cm. What sides do we have to join if we want to obtain a cylinder with a higher volume?

36. We have cut a 8 cm edge cube as shown in the figure. What is the volume of each portion?

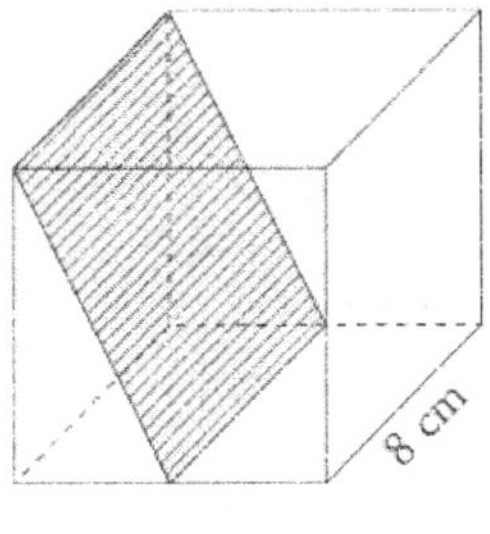

37. When opening a cone we obtain a circle sector with angle 120° and area 84.78 cm^2. Calculate the total area and the volume of the cone.

38. A cylinder and a cone have the same total area, 96π cm^2, and the same radius, 6 cm. Which of them will have a higher volume?

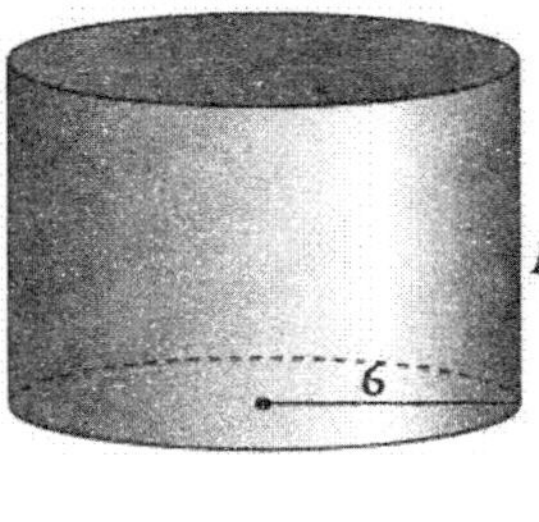

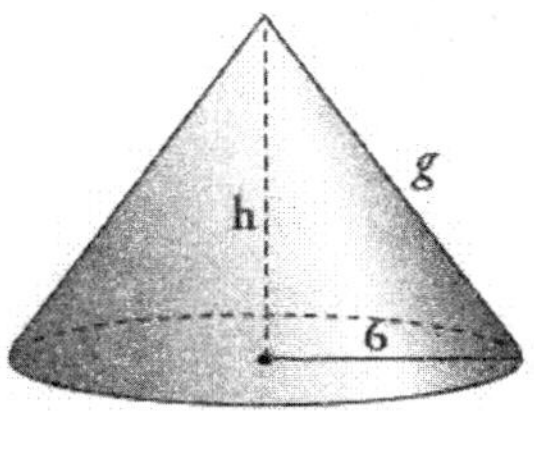

39. a) Which of the following solids are polyhedrons? b) How many of them are regular?

c) Check which of them verify the Euler's formula.

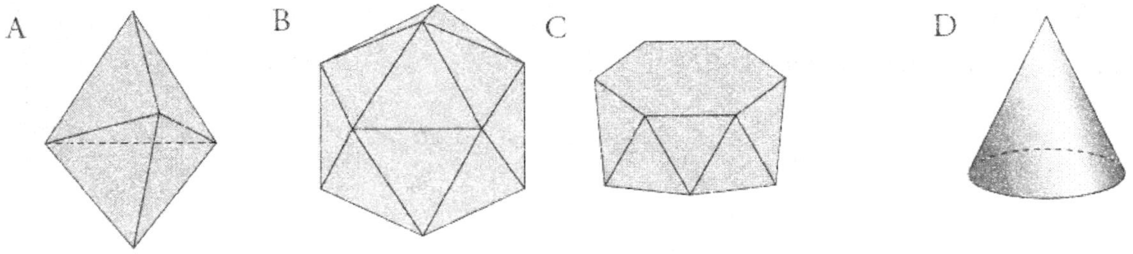

40. If in a cone we reduce into its half the base radius and keep its height, will its volume reduce into its half? And if we keep the base and reduce its height into the half?

41. A square base pyramid is cut by a horizontal plane, parallel to the base and passing by the medium point of the height. What will be relationship between volumes of higher and lower pyramids?

42. A cube and a sphere have the same area. Which of them has a higher volume? Do it by giving a value to the sphere's radius.

43. In a cylinder of diameter identical to the height, we inscribe a sphere. What is the relationship between lateral area of the cylinder and sphere's area?

44. What is the relationship between the volumes of this sphere and this cone?

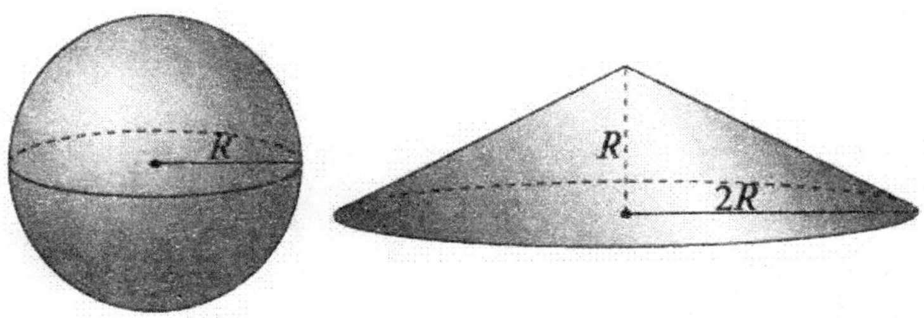

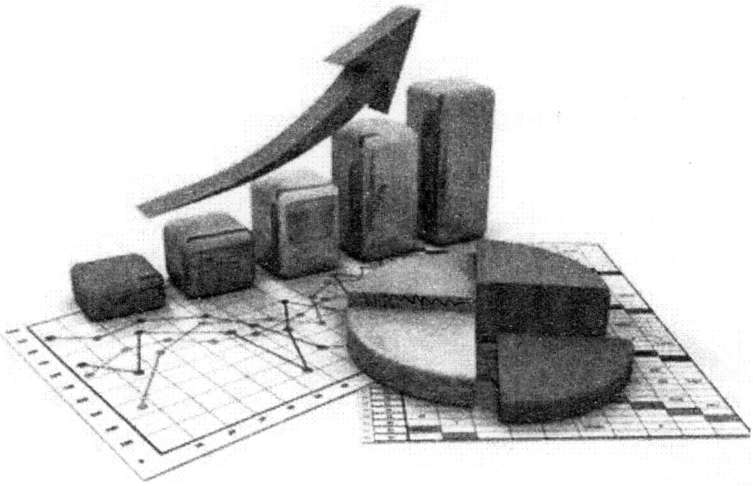

Unit 12.- Statistics

1. Introduction to statistics

Statistics have become an important part of everyday life. We are confronted by them in newspapers and magazines, on television and in general conversations. We encounter them when we discuss the cost of

living, unemployment, medical breakthroughs, weather predictions, sports, politics and the state lottery. Although we are not always aware of it, each of us is an informal statistician. We are constantly gathering, organizing and analyzing information and using this data to make

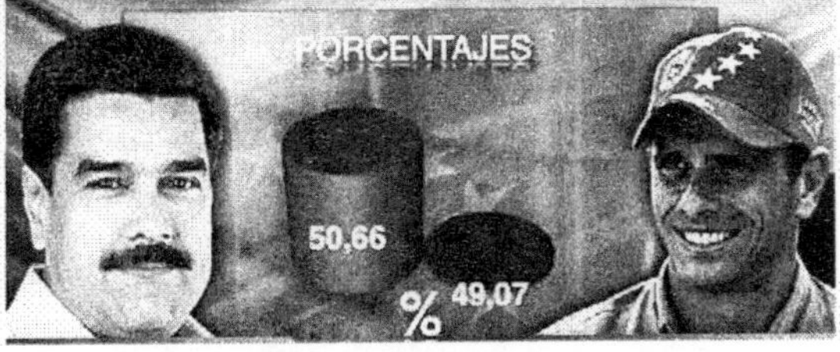

judgments and decisions that will affect our actions. In this unit we will try to improve the students understanding of the elementary topics included in statistics. The unit will begin by discussing terms that are commonly used in statistics. It will then proceed to explain and construct frequency distributions, histograms, frequency polygons and cumulative frequency polygons. Next, the unit will define and compute measures of central tendency including the mean, median and mode of a set of numbers. Measures of dispersion including range and standard deviation will be discussed.

2. Population and sample

Statistics is a branch of mathematics in which groups of measurements or observations are studied. Statistics deals with methods used to collect, organize and analyze numerical facts. Its primary concern is to describe information gathered through observation in an understandable and usable manner.

Throughout the study of statistics certain basic terms appear frequently. Some of the more commonly used terms are defined below.

Statistics work begins when collecting data. As you can imagine, if you want to study heights of citizens in a town, you cannot measure all the *population*. You must choose an appropriate *sample*.

A ***population*** is a complete set of items that is being studied. It includes all members of the set. The set may refer to people, objects or measurements that have a common characteristic.

A ***sample*** is a relatively small group of items selected from a population. If every member of the population has an equal chance of being selected for the sample, it is called a *random sample*.

For example, the owner of a screw factory wants to make a quality control. He picks up 1 out of every 100 produced screws and then he analyses them.

- The *population* is the total number of the screws of the factory.
- The *sample* is 1% of the population.
- The *individuals* are each one of the screws.

3. Statistical variables

Data are numbers or measurements that are collected. Data may include numbers of individuals that make up the census of a city, ages of pupils in a certain class, temperatures in a town during a given period of time, sales made by a company, or test scores made by ninth graders on a standardized test.

Variables are characteristics or attributes that enable us to distinguish one individual from another. They take on different values when different individuals are observed.

Some variables are height, weight, age and price. Variables are the opposite of *constants,* whose values never change.

Statistical variables can be classified in several ways:

Quantitative and qualitative variables

- A ***quantitative or numerical variable*** is that one whose data are numbers. For example, heights, number of books, incomes in €, etc.

- A ***qualitative or categorical variable*** is that one whose data have labels (i.e. words). For example, your favourite musical group, a list of the products bought by different families at a grocery store, eyes colour, etc.

Discrete and continuous variables

Numerical variables can be classified into discrete and continuous variables.

- ***Discrete variables*** are those whose data are whole numbers, and are usually a count of objects. For instance, number of pets in a house, number of children in a family, ... It does not make sense to have 3.5 children.

- ***Continuous variables*** are those that may take any real value. For example, the amount of time a group of children spent watching TV, heights, weights, etc. Your weight may be 74 kg, 74.5 kg or 74.568 kg.

Exercises

1. For each of the following cases, indicate what are the population, the variable and the type of variable.

a) Weight of babies that were born last year in Dublin.

b) Favourite subject for the students in school.

c) Number of pets in French households.

d) Political party that the Spanish electors are going to vote for in the next local elections.

e) Weekly time that students from 12 to 16 spend on reading in Italy.

4. Frequency distributions. Frequency tables

Large groups of data have little value until they have been placed in some kind of order. Usually, measurements are listed in ascending or descending order. Such a group is a *distribution*. A ***frequency distribution*** is a table in which measurements and the *frequency* or total number of times that each item occurs is recorded.

Example: The number of televisions in each house of my street is shown in the frequency table:

Number of TVs	Number of houses
0	1
1	5
2	12
3	9
4	1

a) Calculate the number of houses in my street.

b) Calculate the total of number of televisions in my street.

Solution:

a) The total number of houses is: $1 + 5 + 12 + 9 + 1 =$
$= \boxed{28 \text{ houses}}$.

b) The total number of televisions is: $(0\cdot1) + (1\cdot5) + (2\cdot12) + (3\cdot9) + (4\cdot1) = 0 + 5 + 24 + 27 + 4 =$
$= \boxed{60 \text{ televisions}}$.

Absolute and relative frequencies

The ***absolute frequency*** is the number of times a determined value of a variable occurs.

We will write f_i the absolute frequency of x_i.

We have to know that $f_1 + f_2 + \ldots + f_n = N$, where N is the total number of data. (n is the number of different data).

The ***relative frequency*** is the absolute frequency divided by the total number of observations or total number of data. $h_i = \frac{f_i}{N}$

We will write h_i the relative frequency of x_i.

And we have $h_1 + h_2 + \ldots + h_n = 1$

Exercises

2. In the list below is shown the qualification in Maths of 20 students in a class:

3	3	3	4	5
5	5	5	5	6
6	7	7	7	7
8	9	9	9	9

a) Build a table writing the absolute and relative frequencies of this set of data.

x_i	**Absolute frequency** f_i	**Relative frequency** h_i
...		
...		
...		

b) Calculate the sum of all the absolute frequencies and of all relative frequencies.

Exercises

3. Mrs. Parker asked her students about their favourite subject in the school, and the answers were:

Maths	English	Sciences	English	Maths
History	English	Music	English	Maths
English	Maths	English	History	English
English	English	English	English	Maths
History	Sciences	Maths	Maths	Sciences

Make a table and show the absolute and relative frequencies of these data.

4. In thirty shots, a man makes the following scores:

5 2 2 3 4 4 3 2 0 3 0 3 2 1 5 1 3 1 5 5 2 4 0 0 4 5 4 4 5 5

Summarize the absolute and relative frequencies of the different scores.

5. There are 44 students in a group. Each student plays either hockey or tennis but not both. Complete the table.

	Hockey	**Tennis**	**Total**
Girls	8		20
Boys	18		24

Cumulative absolute and relative frequency

- **Cumulative Absolute Frequency** of a datum x_i is the sum of the absolute frequencies of values less or equal than it. We write F_i.

$$F_i = f_1 + f_2 + ... + f_i$$

- **Cumulative Relative Frequency** of a datum x_i is the sum of the relative frequencies of values less or equal than it. We write H_i. It is the quotient between the cumulative absolute frequency and the total number of data. $H_i = h_1 + h_2 + ... + h_i$

Exercises

6. It is shown the number of brothers and sisters of the 25 students:

0 2 1 0 1 1 1 2 1 3 0 1 1 1 1 0 1 1 2 3 2 1 0 4 2

Complete the table with the different frequencies:

x_i	f_i	F_i	h_i	H_i
...				
...				
...				

7. We asked ~~20~~ 18 students of 2nd E.S.O about the age of their parents. The answers were the following:

40 42 44 47 44 41 43 46 42

46 44 45 47 42 45 40 43 47

Complete a frequency table with the different frequencies.

5. Statistical graphs

Many people find it easier to obtain information from pictures than from written material. Statisticians display mathematical relationships with diagrams and graphs. From these pictures numerical data can be summarized clearly and easily.

Bar diagrams

When the data of a frequency distribution have not been grouped in intervals, they can be represented on a *bar diagram*. A bar diagram illustrates the pattern of a distribution. It clearly shows whether the data are spread out evenly or if they tend to cluster about a point.

To build a bar diagram, list the measurements, from lower to higher, horizontally across the bottom of the graph. On the left side vertically list the frequencies or number of times that the measurements occur. Finally, on each measurement, draw a bar with height equal to its frequency. We draw all the bars with the same width.

Bar diagram is suitable for discrete variable (numerical or categorical).

For example, following figure shows a bar diagram indicating frequencies of vowels in a sentence in English:

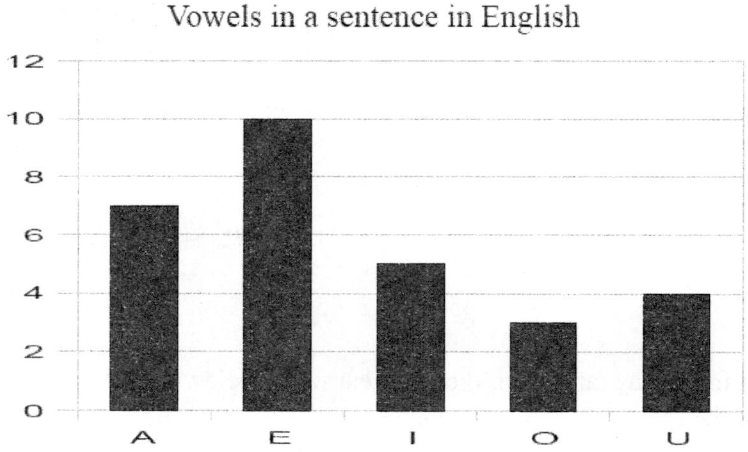

Histograms

The **histogram** is used for variables whose values are numerical and measured on an interval scale (usually for continuous variables).

A histogram divides up the range of possible values in data set into classes or groups. For each group, a rectangle is constructed with an area proportional to the frequency, (if the bars have identical widths, the height of each bar corresponds to the frequency).

Example: Mario decided to collect data about the height of his classmates in the school. These are the data of 40 children (in cm):

163	167	165	159	164	168	161	164	163	164
165	163	167	165	164	164	168	161	164	165
163	164	170	160	157	167	165	172	165	167
164	164	168	161	164	163	164	155	158	162

Solution:

It is useful to make a group frequency table for this case:

Interval	Tally	Frequency
[155-160)	////	4
[160-165)	𝍸 𝍸 𝍸 𝍸 /	21
[165-170)	𝍸 𝍸 ///	13
[170-175)	//	2

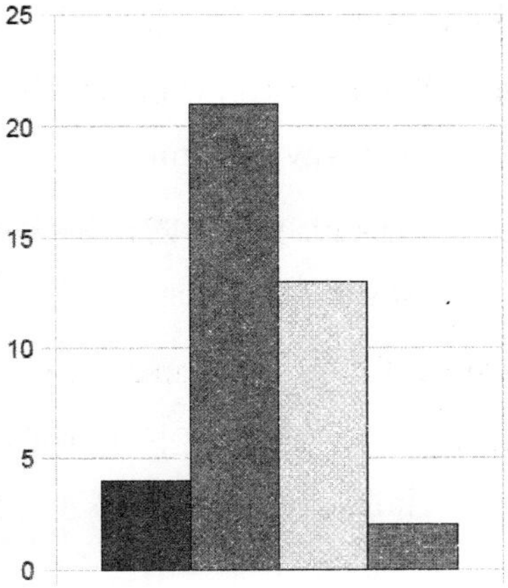

Exercises

8. These are the heights, in cm, of 40 people.

153	134	155	142	140	163	150	135	170	156	171	161	141
153	144	163	140	160	172	157	136	160	134	154	176	154
173	179	160	152	170	148	151	165	138	143	147	144	156
139												

Complete the following frequency table and draw the histogram.

Height (cm)	Number of people
[130,140)	
[140,150)	
[150,160)	
[160,170)	
[170,180)	

9. This frequency table shows the times for 50 runners in a Marathon.

Time (hours)	Number of runners
[1,2)	0
[2,3)	15
[3,4)	23
[4,5)	18
[5,6)	8

Draw a histogram to show the times.

Pie charts

A **pie chart** is a way of summarizing a set of categorical data. It is a circle which is divided into sectors. Each sector represents a particular category. The area of each sector is proportional to the number of cases in that category.

Example: We ask 240 people to name their favourite fruit. With these results, draw a pie chart to illustrate the information.

Fruit	Apple	Banana	Orange	Other
Number of people	50	80	72	38

Solution: First, we calculate the angle for one person: $360^\circ : 240 = 1.5^\circ$

Then, we calculate the angles of each category:

Apple: $50 \cdot 1.5^\circ = 75^\circ$

Banana: $80 \cdot 1.5^\circ = 120^\circ$

Orange: $72 \cdot 1.5^\circ = 108^\circ$

Other: $38 \cdot 1.5^\circ = 57^\circ$

And then, we measure, colour and label the sectors:

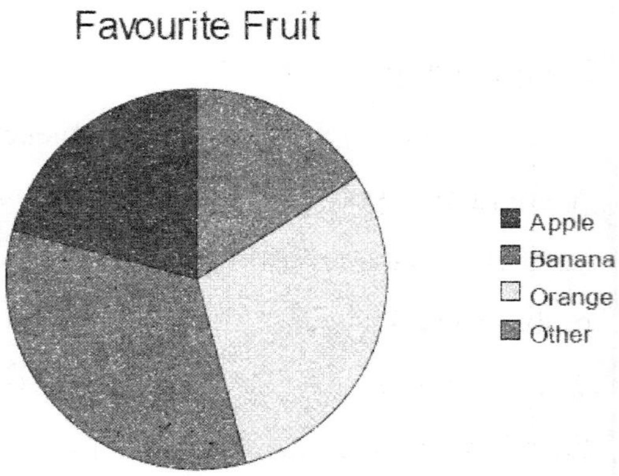

Exercises

10. The Manchester United football team plays 36 matches in a season. They win 15 matches,

and they draw just 8 matches.

a) Calculate the number of matches that they lose.

b) Calculate the angle one match represents in a pie chart.

c) Calculate the angle of each category in the pie chart.

d) Draw a pie chart to show the information.

Exercises

11. The weather record for 60 days is shown in the frequency table.

Weather	Number of days
Sunny	15
Cloudy	18
Rainy	14
Snowy	3
Windy	10

a) Calculate the angle one day represents in a pie chart.

b) Calculate the angle of each category in the pie chart.

c) Draw a pie chart to show the data.

Frequency polygons

A **frequency polygon** is a line graph which can be used to represent the frequency of a set of numbers. It is formed by connecting a series of points. The abscise of each point is the midpoint of the interval in which the point lies. The ordinate of each point is the frequency for the interval. The polygon is closed at each end by drawing a line from the endpoints to the horizontal axis at the midpoint of the next interval.

Example: In an intelligence test, 200 people have been examined. Results are given in the following table.

Interval	f_i
30 – 40	6
40 – 50	18
50 – 60	76
60 – 70	70
70 – 80	22
80 – 90	8

Draw a histogram to represent them and build, also, the frequency polygon.

Solution:

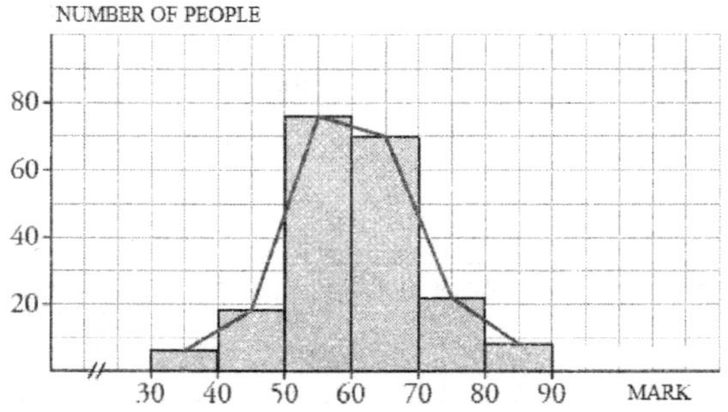

Exercises

12. These are the heights, in cm, of 40 people.

153	134	155	142	140	163	150	135	170	156	171	161	141
153	144	163	140	160	172	157	136	160	134	154	176	154
173	179	160	152	170	148	151	165	138	143	147	144	156
139												

Complete the following frequency table and draw the frequency polygon:

Height (cm)	Number of people
[130,140)	
[140,150)	
[150,160)	
[160,170)	
[170,180)	

6. Measures of central tendency

A ***measure of central tendency*** is a single value that attempts to describe a set of data by identifying the central position in that set of data. As such, measures of central tendency are sometimes called measures of central location. The *mean* (often called the average) is most likely the measure of central tendency that you are most familiar with, but there are others, such as, the *median* and the *mode*.

In the following sections we will look at the mean, mode and median and learn how to calculate them and under what conditions they are most appropriate to be used.

Mean

The ***average*** or ***mean*** of a list of numbers is the total of all values divided by the number of values.

To calculate the mean, we can use the absolute frequencies of values, multiplying every value by its absolute frequency, then adding all these products and finally dividing by the number of values. It is shown in the formula:

$$\bar{x}=\frac{x_1\cdot f_1+x_2\cdot f_2+\ldots+x_n\cdot f_n}{N}=\frac{\sum x_i\cdot f_i}{N}$$

Example: Find the mean of 10, 11, 7 and 8:

Solution: $\bar{x}=\frac{10+11+7+8}{4}=\frac{36}{4}=\boxed{9}.$

Example: Find the mean of the following values:

4 4 4 4 5 5 6 6 6 6

6 7 8 8 8 9 9 9 9 10

Solution: We can use a frequency table:

x_i	f_i
4	4
5	2
6	5
7	1
8	3
9	4
10	1

$$\bar{x}=\frac{4\cdot 4+5\cdot 2+6\cdot 5+7\cdot 1+8\cdot 3+9\cdot 4+10\cdot 1}{20}=\frac{16+10+30+7+24+36+10}{20}=\frac{133}{20}=\boxed{6.65}.$$

Exercises

13. Calculate the mean of: 5, 3, 54, 93, 83, 22, 17 and 19.

14. Calculate the mean of values given as a table:

x_i	f_i
2	5
3	5
4	7
5	8

Median

The **median** of a list of values is found by ordering them from lower to higher. Once they are arranged, median is the central value. Median is the value that has as many values less than it as higher than it.

- If the list has an **odd** number of values, the median is the middle number in this ordering.
- If there is an **even** number of values, the median is the average of the two middle numbers.

Example: Find the median of the following values: a) 8, 7, 5, 6, 9; b) 9, 6, 7, 5, 6 ,7.

Solution:

a) First, we have to order the numbers: 5, 6, 7, 8, 9. As there are five (odd) data, median is 7. **Me = 7**.

b) We order the numbers: 5, 6, 6, 7, 7, 9. There are six (even) data, so, we get the two middle numbers, (6 and 7) and we calculate the mean of both numbers: $Me = \frac{6+7}{2} = 6.5$. Median is 6.5. **Me = 6.5**.

Mode

The **mode** in a list of numbers is the most repeated number (or numbers). It is the value with the highest absolute frequency.

Example: Find the mode of following sets:

a) 2, 3, 3, 6, 4, 3, 2, 5, 6, 3;   b) 2, 2, 2, 3, 4, 4, 4, 5, 6, 6.

Solution:

a) The mode is 3, because it is the number that appears most often in the list. $\boxed{Mo = 3}$.

b) We have two modes in this case, 2 and 4, because they appear three times. Both numbers have the highest absolute frequency. $\boxed{Mo = 2, 4}$.

Exercises

15. The tallest 4 trees in a park have heights in meters of 40, 52, 50 and 55. Find the mean and the median of their heights.

16. Find the mean, median and mode for the following data:

10, 12, 13, 12, 13, 10, 14 and 13.

17. Twenty families are asked about how many children they have. These are the answers:

3 3 4 1 2 3 2 5 1 0

2 2 3 2 4 2 5 3 4 3

a) Complete the table with the frequencies: f_i, h_i, F_i and H_i.

b) Find the mean, median and mode.

7. Dispersion measures

While measures of central tendency are used to estimate "mean" values of a dataset, **measures of dispersion** are important to describe the spread of the data, or its variation around a central value. Two distinct samples may have the same mean or median, but completely different levels of variability, or vice versa.

For example, imagine you can choose working for one of two companies. First thing you are going to ask is about salary. Imagine both companies have the same *mean* salary (for the same hours of work). But, in company A, all the workers earn salaries very close to the mean, while in company B, there are employees who earn the mean salary, others earning much less and others earning much more. What would you decide now?

A proper description of a set of data should include both of these characteristics (central tendency and dispersion measures). There are several dispersion measures that can be used, each with its own set of advantages and disadvantages.

Range

Range is defined as the difference between the largest and smallest sample values.

Range depends only on extreme values and provides no information about how the remaining data is distributed.

Mean deviation

We can easily calculate how much each number deviates from the mean.

Mean deviation is the mean of these deviations. $MD = \frac{|x_1 - \bar{x}| + |x_2 - \bar{x}| + ... + |x_n - \bar{x}|}{n}$

Variance

The Variance (σ^2) is a measure of how spread our numbers are. Its symbol is the squared of Greek letter sigma.

The **variance** is calculated as the average of the **square** of the differences from the mean. Its formula is:

$$\sigma^2 = \frac{(x_1 - \bar{x})^2 + (x_2 - \bar{x})^2 + \ldots.. + (x_n - \bar{x})^2}{n} = \frac{\sum (x_i - \bar{x})^2}{n}$$

There is an equivalent formula to calculate variance. Sometimes it is easier to use:

$$\sigma^2 = \frac{x_1^2 + x_2^2 + \ldots.. + x_n^2}{n} - (\bar{x})^2 = \frac{\sum x_i^2}{n} - (\bar{x})^2$$

Standard deviation

The **standard deviation** is defined as the square root of variance. So, it is denoted by σ.

$$\sigma = \sqrt{\sigma^2}$$

Example: Calculate the dispersion measures of this weights distribution:

83, 65, 75, 72, 70, 80, 75, 90, 68, 72

Solution: First of all, we arrange data: 65, 68, 70, 72, 72, 75, 75, 80, 83, 90.

And, calculate the mean, because we will need it after:

$$\bar{x} = \frac{65+68+70+72+72+75+75+80+83+90}{10} = \frac{750}{10} = 75$$

- Range: 90 – 65 = $\boxed{25}$.

- Mean deviation: $MD = \dfrac{|68-75|+|68-75|+|70-75|+|72-75|+...........+|90-75|}{10} = \dfrac{56}{10} = \boxed{5.6}$.

- Variance: $MD = \dfrac{(65-75)^2+(68-75)^2+(70-75)^2+.........+(90-75)^2}{10} = \dfrac{506}{10} = \boxed{50.6}$.

- Standard deviation: $\sigma = \sqrt{\sigma^2} = \sqrt{50.6} = \boxed{7.11}$.

Variation coefficient

Standard deviation gives us information about dispersion of data. But, sometimes, we want to compare dispersions of different sets of data. If magnitudes of data in different sets of data are very distinct, standard deviation might not be useful.

For example, imagine we have two sets of weights: set A are weights of cars and set B are weights of oranges. Obviously, standard deviation of set B is going to be less than in set A, but that does not mean that in this set there is a lower dispersion of data, it is just because its data a quite lower.

In these cases, we need a new parameter, the *variation coefficient*.

The **variation coefficient** is defined as $VC = \dfrac{\sigma}{\overline{x}}$

As you can imagine, the higher VC is, the higher dispersion is.

Example: In three restaurants, A, B and C, we are investigating salaries (€ per weekend).

	A	B	C
Mean	943	132	37
Standard deviation	148	22	12

In what restaurant do they have the most homogenous salaries?

Solution:

$$VC_A = \frac{943}{148} = 6.37 \qquad VC_B = \frac{132}{22} = 6 \qquad VC_C = \frac{37}{12} = 3.08$$

So, as the lowest VC has been obtained for restaurant C, this is the one having the most homogenous salaries.

Exercises

18. In two groups, 3rd A and 3rd B, we have done the same exam. Results are in the following table.

	Mean	**Std. Dev.**
3rd A	5.8	2.9
3rd B	6.3	1.2

a) Calculate their variation coefficients and indicate in which group results have more dispersion.

b) In a group, there were 6 nines and 5 ones, while in the other one, there were only 1 nine and 2 ones. Could you indicate these groups?

Review exercises

1. Indicate, for each of the five cases proposed:

- What the population is.
- What the variable is.
- Variable type: qualitative or quantitative; quantitative discrete or continuous.

a) Birth weight of babies.

b) Favourite professions for students.

c) Number of pets is houses.

d) Weekly time spent reading by ESO students in Spain.

e) Number of yellow cards in the last football season.

2. The percent of vehicles registered during the month of October 2006 is outlined in this table (data are approximate):

Type of vehicle	Percentage
Cars	69
Lorries	17
Motorbikes	
Busses	0.15
Others	1.45

a) Find the percent of motorcycles registered.

b) Calculate what was the total number of vehicles enrolled, knowing that exactly 279 buses were registered.

c) The set of vehicles registered, is population or sample?

d) Tell what type of variable it is.

3. In a given region, a study on fatal accidents produced at work according to the sector of activity has been done. Here are the results:

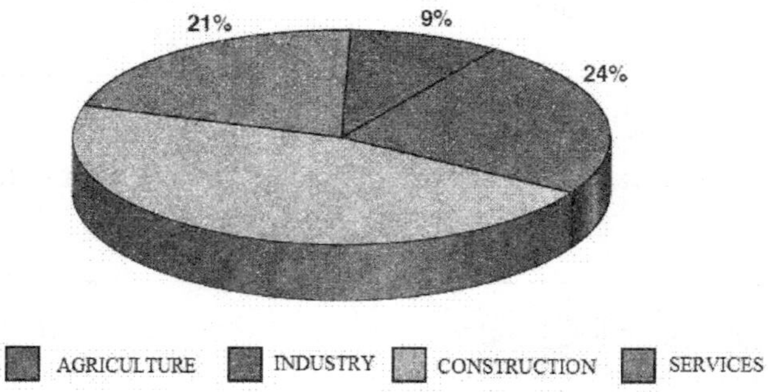

a) What is the percent of fatal accidents in the construction sector?

b) If there were 135 fatal accidents in agriculture, what was the total number of fatal accidents in the region?

c) How many fatalities occurred in each of the sectors?

4. Figure 1 shows evolution of unemployed ratio in Spain from 1995 to 2000 year. Figure 2 shows a table for the same data, in which only difference is scale for Y-axis.

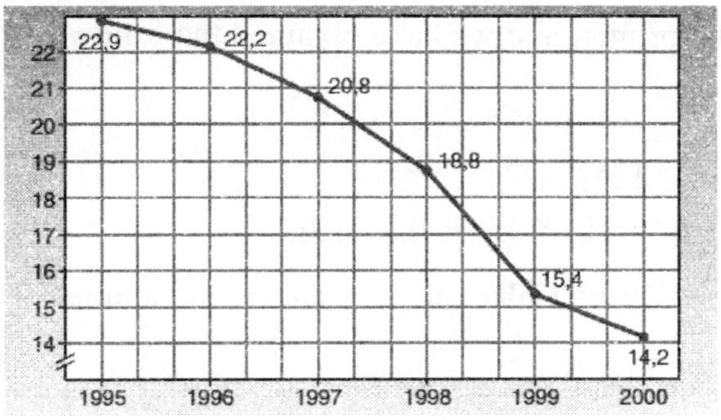

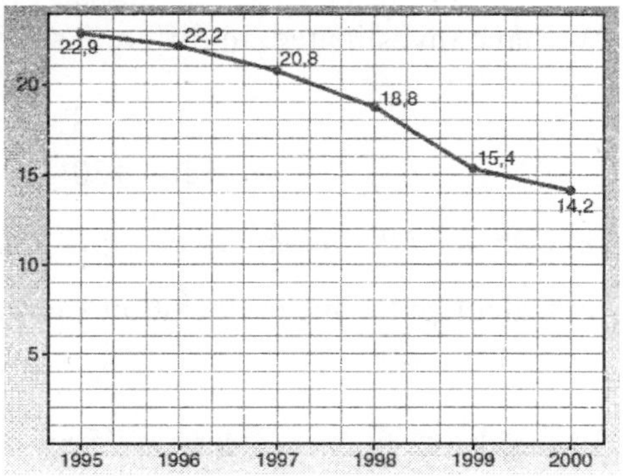

a) Do both of them give the same sensation?

b) Which one do you think a government would choose to give unemployment data and which one would be chosen by opposition party?

Tables and graphs elaboration

5. When being asked about the number of books they have read for last month, 3rd ESO students answers were the following:

2 1 3 1 1 5 1 2 4 3

1 0 2 4 1 0 2 1 2 1

3 2 2 1 2 3 1 2 0 2

a) Build a frequency table.

b) Draw the corresponding bar diagram.

6. Chosen colour by Spanish people when buying a car is shown in the following table. Draw the corresponding pie chart.

Colour	**Percentage**
Argent grey	36%
Black	22%
Blue	18%
Red	10%
White	8%
Green	4%
Others	2%

Statistical parameters

7. We have investigated, in different shops, price of a printer, and we have obtained the following results:

146 - 150 - 141 - 143 - 139 - 144 - 133 - 153

a) Calculate the mean price.

b) What is the median?

c) Calculate the mean deviation and the range.

d) Calculate the standard deviation.

8. In family A, father's salary is 950 €, and mother's one is 1600 €. In family B, the father earns 1800 €/month, and the mother, 750 €.

a) What is the mean salary of each family?

b) In what family is there more dispersion? What is the range for each family?

9. Counting the number of misprints in a book, Peter has obtained the following data:

Number of misprints	0	1	2	3	4	5
Number of pages	50	40	16	9	3	2

Calculate the a) mean, b) standard deviation and c) the mode.

10. In a road speed control, they have obtained the following data:

Speed (km/h)	**Number of cars**
60 – 70	5
70 – 80	15
80 – 90	27
90 – 100	38
100 – 110	23
110 – 120	17

a) Build a table containing class marks and frequencies.

(*<u>Clue:</u> class mark of interval 50-60 is 55*).

b) Calculate the mode and the standard deviation.

c) What percentage has speed higher than 90 km/h?

11. Teresa and Rosa are basketball players. Their punctuations for a week training have been:

Teresa	16	25	20	24	22	29	18
Rosa	23	24	22	25	21	20	19

a) Calculate the mean punctuation for each one.

b) Calculate their standard deviations and variation coefficients. Who is more regular?

12. 40 people were asked about the number of people living at home:

4 5 3 6 3 5 4 6 3 2   2 4 6 3 5 3 4 5 3 6

4 5 7 4 6 2 3 4 4 3   4 4 5 3 2 6 3 7 4 3

a) Build a frequency table and draw the corresponding diagram.

b) Calculate the mean, median, mode and standard deviation.

13. When measuring the birth weight of a certain animal species, we have obtained the following data:

Weight (kg)	**Number of animals**
3.5 – 4.5	1
4.5 – 5.5	8
5.5 – 6.5	28
6.5 – 7.5	26
7.5 – 8.5	16
8.5 – 9.5	1

a) Represent these data in the appropriated graph.

b) Calculate the mean and standard deviation.

c) What percentage of animal had a birth weight between 5.5 and 6.5 kg? And between 4.5 and 8.5 kg?

14. These are the weekly study hours of a group of students:

14	9	9	20	18	12	14	6	14	8
15	10	18	20	2	7	18	8	12	10
20	16	18	15	24	10	12	25	24	17
10	4	8	20	10	12	16	5	4	13

a) Organize these data in a table with the following intervals:

1.5-6.5; 6.5-11.5; 11.5-16.5; 16.5-21.5; 21.5-26.5

b) Build its frequency table and histogram.

c) Calculate the mean and standard deviation.

15. Monthly expenses of a company A have mean 60 000 € and standard deviation of 7500 €. In other company B, mean is 9000 €, and standard deviation is 1500 €. Calculate, by using the variation coefficient, which of them has a higher relative variation.

16. Number of words in each sentence of an economy article is:

17	40	22	25	43	21	17	25	37	12
9	37	32	35	30	21	13	27	41	45
36	40	30	48	45	41	39	39	40	38
28	7	33	35	22	34	23			

a) Build a frequency table grouping data into the following intervals:

7 - 13, 14 - 20, 21 - 27, 28 - 34, 35 - 41, 42 - 48

Represent these data in a histogram.

c) Calculate the mean and standard deviation.

17. In my instituto there are 200 students. They were asked about their favorite pet. See the figure and answer the questions:

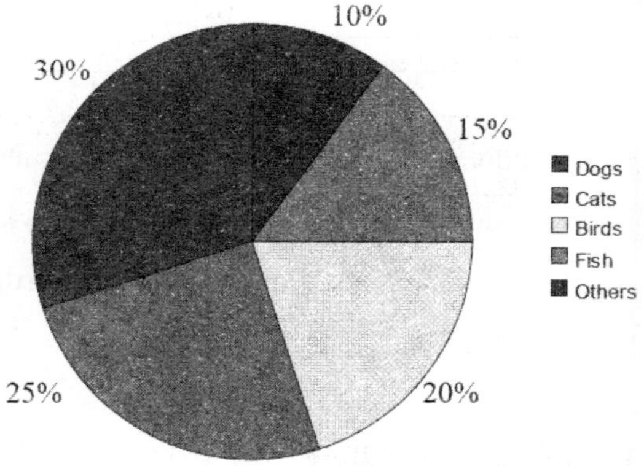

a) How many people did answer "dogs"?

b) And "cats"?

18. Teachers of my instituto are asked about their favorite film genres. The answers were:

Comedy: 27% Action: 18%

Romance: 14% Drama: 14%

Horror: 11% Foreign: 8%

Fiction: 8%

a) Draw a bar chart and a pie chart with the data.

b) Imagine there are 120 teachers in this high school. How many of them did answer "Romance"?

19. These are the number of photographic cameras sold the last year in a store:

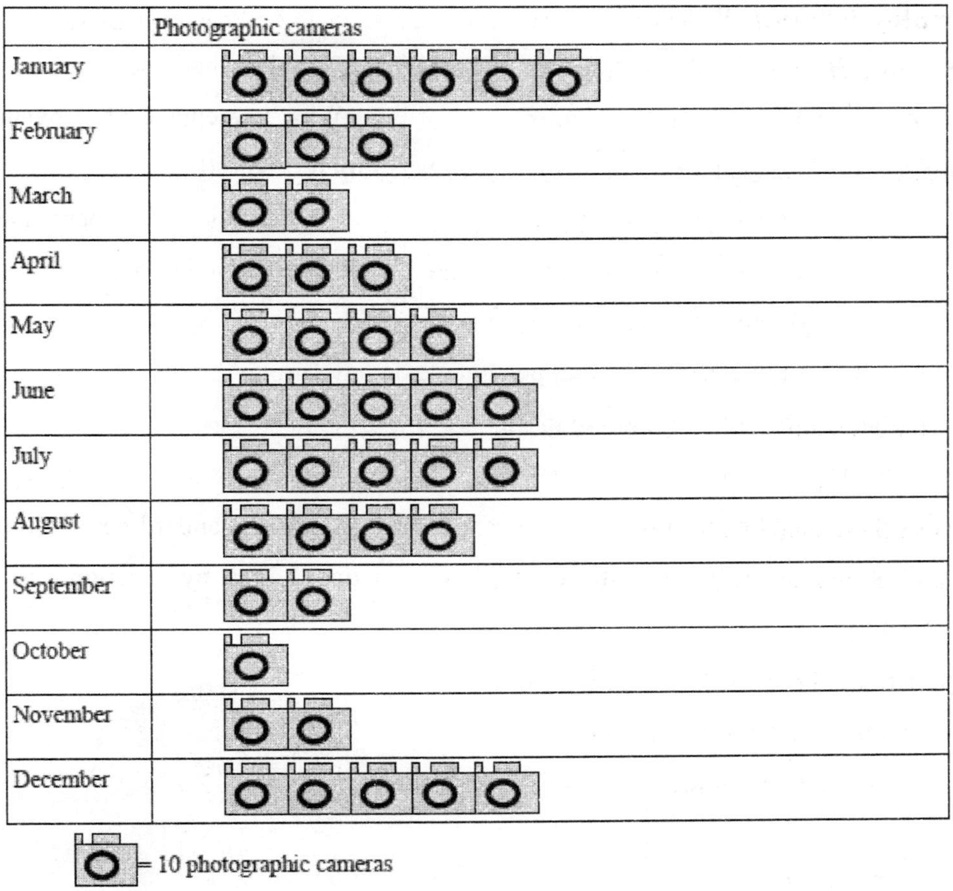

a) How many cameras have been sold every month? Which one was the best month for the salesman?

b) Why do you think was this month? What is the average of sold cameras per week?

20. These are the numbers of cars that the different families of a street have:

1 2 1 1 0 2 3 1 1 2

3 1 0 2 1 2 1 1 2 1

a) Complete the table with the frequencies: f_i, h_i, F_i and H_i.

b) Represent the data using a pie chart and a bar graph.

c) Find the mean, median and mode of the data.

21. Blood group (A, B, AB, 0) has been studied in 200 people, as well as their Rh. Some of the results are in the following table.

	A group	B group	AB group	0 group	Total
Rh+	74		6	70	162
Rh-		3	1		
Total				86	200

a) Complete the table.

b) What percentage of population has group B with Rh+?

c) What is the percentage of population that has Rh-?

d) Between people having group Rh+, what percentage do have group A?

e) Draw a pie chart for different blood groups.

22. These three distributions have the same mean, what is it? Their standard deviations are 3.8, 1.3 and 2.9. Join each distribution with its standard deviation. (Do it by observing the graphs, without calculating).

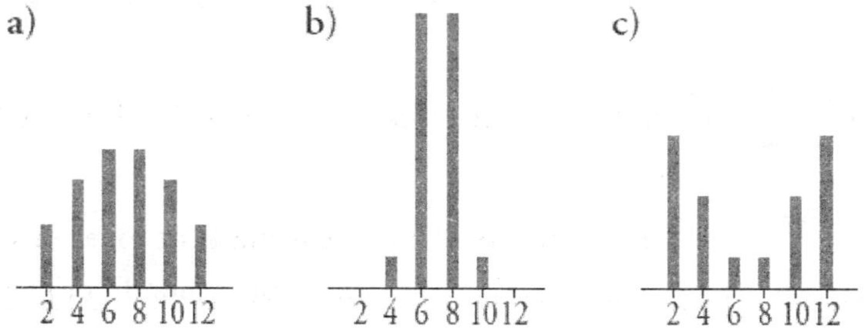

23. A test with 25 questions has been answered by 120 students. Following table shows the results:

Number of correct answers	Percentage
5	10%
15	45%
20	25%
25	

a) Calculate the number of students that answered correctly all the questions.

b) Calculate the mean of corrects answers of population.

c) Calculate the standard deviation.

24. What will happen to mean and standard deviation if we add a same number to all data? And if we multiply them by a same number? Check it with these data: 3, 5, 6, 3, 4, 2, 3, 6.

25. If two distributions have the same mean and first one has a higher standard deviation, which of them will have a higher deviation coefficient?

26. If two distributions have the same standard deviation and first one has a higher mean, which of them will have a higher deviation coefficient?

27. Mean of marks in 3rd A was 6.2, and in 3rd B, 4. If there are 15 students in 3rd A and 35 in 3rd B, what would the mean be if calculated for both groups together?

28. Complete the following table, knowing it mean is 2.7.

x_i	f_i
1	3
2	…
3	7
4	5

29. My mark is calculated as the mean of four exams. If in the three first I have a mean of 4.2, what should be my mark at the fourth exam to pass the subject?

30. My final mark in a subject is calculated from marks of three partial exams: second exam's mark has double value, and third one's value is three times the first one.

a) If my marks have been 5, 6 and 4, What will be my final mark?

b) What would my final mark be if each exam had an importance of 10%, 40% and 50%, respectively?

31. In a company there are 34 employees and 6 directives. Mean salary of all them is 909 €. What is the mean salary of directives if we know that mean of salaries of the rest of workers is 780 €?

Unit 13.- Probability

1. Introduction to probability. Experiments. Sample space. Events

Introduction

Humans often think in probabilistic terms, whether we are conscious of it or not. That is, we decide to cross the street when the probability of being run over by a car is sufficiently low, we go fishing at the lakes where the probability of catching something is sufficiently high, and so on. So, even when people are wholly unfamiliar with the mathematical formalization of probability, there is an inclination to frame uncertain future events in such terms.

Deterministic and random experiments

In our life, in Nature, we can have two types of experiences or, in probabilistic terms, two types of experiments can be done: *deterministic* and *random* experiments. A **deterministic** experience or experiment is that one in which you can know the result before performing it. For example, if today it is Monday, we are *sure* tomorrow it will be Tuesday. Or, if I hold a pencil in my hand and I open my fingers, I am sure, even if I do not do it, pencil will fall, it won't go up. So, deterministic experiences are not in the field of study of probability, because they produce *sure* results.

On the other hand, **random** experiments refer to those experiments whose results cannot be determined before performing them. For example, if I throw a coin on a table, I can have two results, and I cannot determine it if I do not throw it. As we cannot be sure about the result of random experiments, we speak about *probability* of a result to happen.

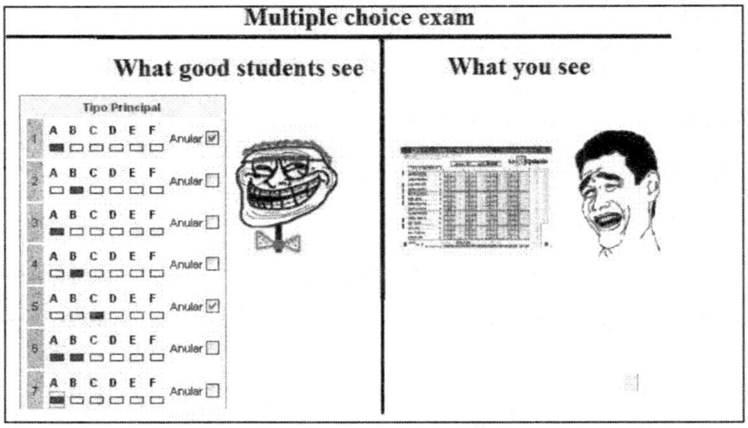

Probability is the branch of Mathematics that studies and quantifies the chance of random experiences results.

Sample set. Elements

Each possible simple result of a random experiment is named ***element***. The ***sample set*** of a random experiment includes all the elements in it.

Example: In a box, there are 3 green balls, 2 red ones and 6 blue. Write its sample set. How many elements does it have?

<u>Solution:</u> Sample set: {G, R, B}. Three elements.

Example: In a bag, there are 10 numbered cards, from 0 to 9. Write its sample set. How many elements does it have?

Solution: Sample set: {0, 1, 2, 3, 4, 5, 6, 7, 8, 9}. Ten elements.

Example: Two dice are thrown on a table. We study the sum of them. Write its sample set. How many elements does it have?

Solution: Sample set: {2, 3, 4, 5, 6, 7, 8, 9, 10, 11, 12}. Eleven elements.

"Dice" is the plural of *"die"*. So, in a game, you can throw one *die* or some *dice*.

Events

An ***event*** is a sub-set of the sample set of a random experiment.

Example: We throw a die on a table. As you already know, its sample set is {1, 2, 3, 4, 5, 6}. Describe the following events:

A = Obtain an even number, B = obtain a multiple of three,

C = obtain a number higher than 3, D = obtain a number equal or higher than 3,

E = obtain a number between 1 and 6 , both included, F = obtain number 8.

Solution: A = {2, 4, 6} B = {3, 6} C = {4, 5, 6} D = {3, 4, 5, 6}

E = {1, 2, 3, 4, 5, 6} F = Ø (empty set).

There are several special events we must consider:

- ***Sure event*** is that one that is always going to happen. Its probability is 100%.
- ***Impossible event*** is that one that is never going to happen. Its probability is 0.
- Given an event A, we can define its ***complementary event***, denoted by $\overline{A}$, as a new event that includes the elements of sample set not included in A. Notice that A + $\overline{A}$ = sample set.

Example: In a box, there are balls numbered from 1 to 10. Describe the following events:

a) A = Obtain a multiple of 4

b) B = Obtain a number from 1 to 10

c) C = Obtain number 20

d) D = Obtain a divisor of 8

e) $\overline{D}$

f) How could you define elements in $\overline{D}$?

Solution:

a) A = {4, 8}

b) B = {1, 2, 3, 4, 5, 6, 7, 8, 9, 10} (sure event)

c) C = Ø (impossible event).

d) D = {1, 2, 4, 8}

e) E = {3, 5, 6, 7, 9, 10}

f) Numbers from 1 to 10 not being divisors of 8.

Intersection and union of two events

- Given two events, A and B, ***intersection of A and B***, denoted by $A \cap B$, is a new event containing elements that are included, at the same time, in A **and** B.
- Given two events, A and B, ***union of A and B***, denoted by $A \cup B$, is a new event containing elements that are included in A **or** B.

Example: Consider the following events of the experiment in which we take a card from a box containing numbered cards from 1 to 10:

A = Obtain an odd number

B = Obtain a number equal or higher than five.

a) Describe events A and B.

b) Describe complementary event of A, $\overline{A}$

c) Describe intersection $A \cap B$

d) Describe union $A \cup B$

Solution:

a) A = {1, 3, 5, 7, 9} B = {5, 6, 7, 8, 9, 10}

b) $\overline{A}$ = {2, 4, 6, 8, 10}

c) $A \cap B$ = {5, 7, 9}

d) $A \cup B$ = {1, 3, 5, 6, 7, 8, 9, 10}

Exercises

1. Classify the following experiments into deterministic and random experiments:

a) Extract a card from a cards game.

b) Check the weight of a liter of mercury.

c) Ask a number to a friend.

d) Throwing three dice and write down their sum.

e) Subtracting two known numbers.

2. Write all the possible results of throwing two coins on a table.

3. We throw a coin and a die. Write the sample set.

4. Given events A = {1, 2, 3} and B = {1, 3, 5}, calculate their union and intersection.

5. When throwing a die with 8 faces, we consider following events:

A = {2, 4, 5, 8} and B = {1, 2, 3, 7}. Calculate:

a) $A \cup B$ b) $A \cap B$ c) $\overline{A \cap B}$ d) $\overline{A} \cup \overline{B}$ e) $\overline{A \cup B}$ f) $\overline{A} \cap \overline{B}$

Could you extract a rule by observing results c) and d)? And e) and f)?

2. Laplace's Law

As we have already said, probability field is that one in which there are not sureness when performing an experience. So, we can only speak about the probability of an event to happen. In this part of the unit, we are going to learn to quantify this probability.

Expression to quantity probability was developed by Laplace. That is the reason by which it is known as **Laplace's Law.**

Probability of an event is calculated by counting favourable elements for the event and possible elements of the experiment, and calculating their quotient.

$$P(A) = \frac{Number\ of\ \ favourable\ elements\ to\ A}{Number\ of\ \ possible\ elements}$$ ***(Laplace's Law).***

Example: In a bag, there are ten numbered balls from 1 to 10. A ball is extracted from the bag. Calculate the following possibilities:

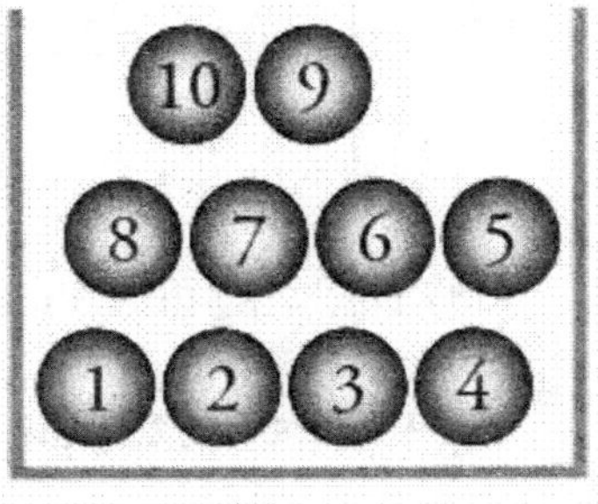

a) Obtaining number 8 b) Obtaining an even number

c) Obtaining a multiple of 5

<u>Solution:</u> In all cases, there are 10 possible elements, 10 balls.

a) There is 1 ball with number 8, so, $P(8) = \boxed{\frac{1}{10}}$

b) There are 5 even numbers between 1 and 10 (2, 4, 6, 8 and 10), so, $P(even) = \frac{5}{10} = \boxed{\frac{1}{2}}$

c) There are 2 multiples of 5 between 1 and 10 (5 and 10), so, $P(even) = \frac{2}{10} = \boxed{\frac{1}{5}}$

Exercises

6. When throwing a die, calculate the probability of obtaining:

a) a multiple of 5 b) a divisor of 2 c) a prime number d) Number 3

e) a divisor of 6 f) an even number divisor of 4 g) multiple of 7

h) less than 10 i) odd number

7. From a pack of Spanish cards we get a card. a) What is the probability of getting a horse? b) And a figure? c) And a "gold"?

8. In a box there are 5 yellow balls and 7 red ones. What is the probability of getting a yellow one? And a green one?

Exercises

9. A card has a black side and white one. We have thrown this card 85 times and we have obtained the white side 43 times. What do you think probability of getting black one is?

10. We throw two dice and sum their punctuations. What is the probability of obtaining:

a) 3 b) Higher than 10 c) 7 d) 4 or 5.

11. A box contains 4 white balls, 2 red ones and 5 black ones. Calculate the probability to obtain:

a) a white ball b) a red ball c) a white or black ball.

12. In a bag, there are 90 balls, numbered from 1 to 90. We extract one of them.

a) What is the probability of getting ball number 59?

b) What is the probability of getting a multiple of 10?

13. In a bag there are balls having two sizes, large (L) and small (S). We extract one of them, write the size (L or S), insert it back to the bag, extract another one, etc. In this way, we observe 84 L balls and 36 S ones. What values would you give to P(L) and P(S)?

14. In a summer school, there are 32 European children, 13 American, 15 African and 23 Asian ones. Representative is randomly chosen. What is the probability of being an European one?

15. Two dice are thrown on a table and sum of punctuations are written. Calculate the probabilities of each sum.

Review exercises

Sample set. Experiments. Events

1. Classify the following experiments into deterministic and random:

a) Extract a card from a pack of cards.

b) Measuring the hypotenuse of a right triangle whose legs are 3 cm and 4 cm.

c) Pushing "ON" in a lamp.

d) Measuring the height of a classroom.

e) Throwing a stone and measuring gravity constant.

f) Finding out the result of a football match before being played.

2. Write the sample set of the following random experiments:

a) Extract a card from a pack of Spanish cards.

b) Extracting a ball from a box containing 5 red, 3 blue and 2 green ones.

c) Throwing two dice and subtracting the results.

d) Throwing two dice and multiplying the results.

3. We throw a dodecahedron shaped die with faces numbered from 1 to 12.

a) What is the sample set?

b) Write the events: A = less than 5; B = more than 4;

C = even number; D = Not a multiple of 3.

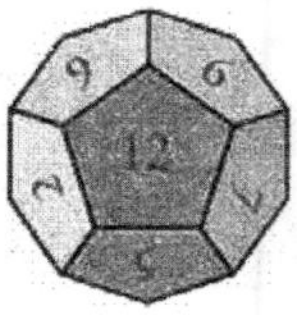

4. We write each letter of the word GAME in a different card, insert them into a bag and extract one.

a) Write the sample set. b) Describe the event "obtaining a vowel".

c) If the word was PROBABILITY, how would you answer?

5. Consider the experiment in which we randomly get a domino token.

a) Describe the event A = Choosing a token whose numbers' sum is 6.

b) Describe the event B = Choosing a token whose numbers' product is 12.

c) Calculate the union and intersection of A and B.

6. In a box there are 15 numbered balls from 1 to 15 and we extract one of them. Write the elements of the following events:

a) Multiple of 3 b) Multiple of 2 c) Higher than 4

d) higher than 3 and lower than 8 e) Odd number.

7. When throwing a die with 6 faces, A = {2, 4} and B = {1, 2, 3}. Calculate:

a) $A \cap B$ b) $A \cup B$ c) The complementary of events A, B, $A \cap B$ and $A \cup B$.

8. We throw a die with 6 faces and consider events A = {1, 3, 5, 6}, B = {1, 2, 4, 5} and C = {3, 4}. Calculate: a) $\overline{A}$ b) $\overline{B}$ c) $\overline{C}$ d) $A \cup B$ e) $A \cap B$

f) $A \cup B$ g) $\overline{A \cup B}$ h) $\overline{A} \cap \overline{B}$ i) $\overline{A} \cup \overline{B}$

9. Calculate the probability of obtaining a 5 when throwing each of the following dice:

a)

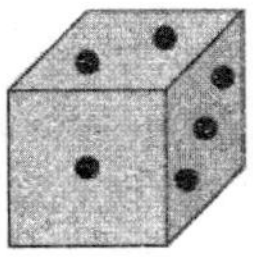

b)

c)

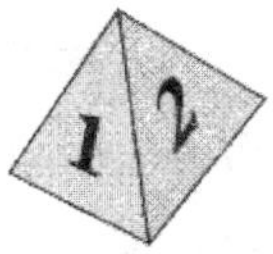

10. In a bag there are 6 red bolls, 4 blue, 7 green, 2 yellow and 1 black one. We randomly extract one of them. Calculate the probability of obtaining one being:

a) Blue b) Not black c) Red or green d) Nor yellow neither black.

11. In a test exam there are 80 units, from which you choose one randomly. A person has studied 60 of them. Calculate the probability of a) passing and b) failing the exam.

12. Calculate the probabilities of the following events when throwing a correct die:

a) Multiple of 3 b) Multiple of 2 c) Higher than 1 d) Lower than 5

e) Lower than 1

13. In a school, there are 990 students, being 510 of them, girls. If we randomly choose a student, what is the probability of choosing a boy?

14. In a high school, students are distributed as indicated in the table:

1st ESO	2nd ESO	3rd ESO	4th ESO	1st BACH	2nd BACH
210	250	260	220	140	120

If we randomly choose a student, what is the probability that he/she studies:

a) 3rd ESO b) ESO c) Bachiller

15. In a book having 120 pages, we have counted the number of misprints in each page. Results are in the following table. When randomly chosen a page of this book:

Number of errors	Number of pages
0	58
1	42
2	16
3	3
4	1

a) What is the probability of choosing a page without misprints?

b) What is the probability of choosing a page having exactly 2 misprints?

c) What is the probability of choosing a page having misprint/s? And having more than 3?

16. We have a card having a white and a black faces. We throw it 3 times. What is the probability of obtaining the black face a) exactly once? b) twice and b) three times?

17. We randomly extract a domino token. Calculate the probability of getting:

a) A sum of punctuations being lower than 4.

b) A sum of punctuations being multiple of 3.

c) A "Double" token (having both identical punctuations).

18. We throw two dice: Calculate the probability of obtaining

a) A product of punctuations of 5. b) A product of punctuations of 6.

c) A product of punctuations of 4. *(Clue: build a table with possible results).*

19. We throw 100 times a tetrahedron shaped die (faces: 1, 2, 3 and 4), writing the number of the hidden face. We have obtained the following results:

Face	1	2	3	4
Frequency	28	22	30	20

Calculate the probabilities of the following events:

a) Multiple of 3 b) Multiple of 2 c) Higher than 1 d) Lower than 1.

20. Probability of an event is 0.2. What is the probability of its complementary?

21. If in a die, P (1) = P (2) = P (3) = 0,14 y P (4) = P (5) = P (6) = x, what s the value of x?

22. In a non-correct die, probabilities of each face are the following. Knowing that P(4) = 2·P(5), what are the values of *a* and *b*?

Face	1	2	3	4	5	6
P(i)	0.1	0.1	0.1	a	b	0.4

23. We randomly choose a number from 1 to 30. Consider the events A = "Obtaining an even number being lees or equal to 14", B = "Obtaining a multiple of 3 less or equal to 10" and C = "Obtaining a multiple of 10". Calculate the probabilities of:

a) $A \cup B$ b) $A \cup C$ c) $A \cup \overline{B}$ d) $C \cup B$ e) $B \cap C$ f) $\overline{A} \cap B$

24. Consider a game in which you throw two dice and you win if you obtain a sum of punctuations of 11 or 7. What is your probability of winning?

25. Numbers in a bingo's roulette are from 1 to 36. Calculate the following probabilities:

a) Getting number 9. b) Getting an even number.

c) Getting a number finishing in a 3.

d) Getting a multiple of 4

e) Getting a number having two identical digits

f) Getting a number starting with a 4.

ADVANCED

26. A bottle contains 20 balls, between black, red and green. We do not know how many there are with each colour, because the bottle is not transparent. We can only see into the bottle in a very little transparent part at the top of the bottle, where there is exactly the space for one ball. We can rotate the bottle down and see this ball. We have moved and rotated the bottle for 1000 times, obtaining the following frequencies:

f(Black) = 461   f(Red) = 343   f(Green) = 196

We can estimate the number of balls of each colour. We are doing it for you with black ones.

$$\text{f(Black)} = 461 \quad \rightarrow \quad \text{h(Black)} = \frac{461}{1000} = \boxed{0.461}$$

There are 20 balls. Let's suppose there are "b" black balls. Probability of seeing a black ball will be:

$$\text{P(Black)} = \frac{n}{20}$$

By equalizing to the corresponding frequency, $\frac{n}{20} = 0.461 \rightarrow n = 20 \cdot 0.461 = 9.22$.

So, we can estimate that there are $\boxed{\text{9 black balls}}$.

Now, do it for the others.

27. Students in a group are distributed as follows:

	Girls	Boys	Totals
Using glasses	**3**	**6**	9
Not using glasses	**12**	**10**	21
Totals	15	16	30

We randomly choose a person. Calculate the probability of being:

a) A girl   b) A person who uses glasses   c) A girl who uses glasses

d) Knowing it is a person who uses glasses, probability of being a boy

e) Knowing it is a girl, probability of using glasses.

28. In a restaurant there are 28 men and 32 women. 16 men and 20 women have had meat for lunch, and the others, fish. If we randomly choose a person, calculate the probability of the following events:

a) Being man b) Having had fish c) Being a man and having had fish

d) Knowing he/she has had fish, probability of being a man.

29. In a school, there are 20 boys and 16 girls. Half of the boys and the three fourths of the girls are brown, and the others, blonde. What is the probability that when randomly chosen a person, he/she is a boy or had brown hair?

30. In a town, newspaper A is read by 30% of citizens, and newspaper B is read by 20% of citizens, while 7% of the citizens read both of them. If we randomly choose a citizen, what is the probability that he:

a) Reads some of the newspapers? b) Does not read any newspaper.

c) Read exactly one newspaper.

31. Luis and Juan must clean their room. Luis puts in a bag 3 red balls, 2 green and 1 blue, and purposes to his brother to get one of them. If it is red, Juan will clean, and if it is blue, Luis will clean.

a) Is this a neutral system?

b) Juan does not accept this system and he purposes another one to his brother: if he gets a red ball, he will clean, and if it is blue or green, Luis will clean. Is this a neutral system?